SpringerBriefs in Applied Sciences and Technology

Series editor

Janusz Kacprzyk, Polish Academy of Sciences, Systems Research Institute, Warsaw, Poland

SpringerBriefs present concise summaries of cutting-edge research and practical applications across a wide spectrum of fields. Featuring compact volumes of 50–125 pages, the series covers a range of content from professional to academic.

Typical publications can be:

- A timely report of state-of-the art methods
- An introduction to or a manual for the application of mathematical or computer techniques
- A bridge between new research results, as published in journal articles
- A snapshot of a hot or emerging topic
- An in-depth case study
- A presentation of core concepts that students must understand in order to make independent contributions

SpringerBriefs are characterized by fast, global electronic dissemination, standard publishing contracts, standardized manuscript preparation and formatting guidelines, and expedited production schedules.

On the one hand, **SpringerBriefs in Applied Sciences and Technology** are devoted to the publication of fundamentals and applications within the different classical engineering disciplines as well as in interdisciplinary fields that recently emerged between these areas. On the other hand, as the boundary separating fundamental research and applied technology is more and more dissolving, this series is particularly open to trans-disciplinary topics between fundamental science and engineering.

Indexed by EI-Compendex and Springerlink.

More information about this series at http://www.springer.com/series/8884

Francisco Chinesta
Emmanuelle Abisset-Chavanne

A Journey Around the Different Scales Involved in the Description of Matter and Complex Systems

A Brief Overview with Special Emphasis on Kinetic Theory Approaches

Francisco Chinesta
ENSAM ParisTech
Paris
France

Emmanuelle Abisset-Chavanne
École Centrale de Nantes
Nantes
France

ISSN 2191-530X ISSN 2191-5318 (electronic)
SpringerBriefs in Applied Sciences and Technology
ISBN 978-3-319-70000-7 ISBN 978-3-319-70001-4 (eBook)
https://doi.org/10.1007/978-3-319-70001-4

Library of Congress Control Number: 2017958588

© The Author(s) 2018
This work is subject to copyright. All rights are reserved by the Publisher, whether the whole or part of the material is concerned, specifically the rights of translation, reprinting, reuse of illustrations, recitation, broadcasting, reproduction on microfilms or in any other physical way, and transmission or information storage and retrieval, electronic adaptation, computer software, or by similar or dissimilar methodology now known or hereafter developed.
The use of general descriptive names, registered names, trademarks, service marks, etc. in this publication does not imply, even in the absence of a specific statement, that such names are exempt from the relevant protective laws and regulations and therefore free for general use.
The publisher, the authors and the editors are safe to assume that the advice and information in this book are believed to be true and accurate at the date of publication. Neither the publisher nor the authors or the editors give a warranty, express or implied, with respect to the material contained herein or for any errors or omissions that may have been made. The publisher remains neutral with regard to jurisdictional claims in published maps and institutional affiliations.

Printed on acid-free paper

This Springer imprint is published by Springer Nature
The registered company is Springer International Publishing AG
The registered company address is: Gewerbestrasse 11, 6330 Cham, Switzerland

To our families,
Ofelia, Nathalie, Lucia & Daphné (F. Ch.)
Robin, Alice & Simon (E. A-C.)

Acknowledgements

The authors wish to acknowledge the support of the ESI Group through its Chair at Ecole Centrale de Nantes (France), the Institut Universitaire de France, and the Spanish Royal Academy of Engineering. Moreover, they acknowledge the contribution of several colleagues from École Centrale de Nantes, Université catholique de Louvain (Prof. Roland Keunings), ENSAM Angers (Prof. Amine Ammar), and the University of Zaragoza (Prof. Elias Cueto). The authors also wish to acknowledge the detailed reading of the whole text by Adrien Scheuer (ECN & UCL).

Contents

1 The Schrödinger Equation 1
1.1 The History of Quantum Mechanics in Small Bits 1
1.2 Planck Versus the Ultraviolet Catastrophe 4
1.2.1 Rayleigh–Jeans Theory 4
1.2.2 Planck Theory 6
1.3 An Intuitive Approach to the Schrödinger Equation 7
1.4 The Feynman Approach 10
1.5 The Schrödinger Equation 12
1.6 Relations Between Position and Momentum Wavefunctions 16
1.6.1 Calculating Expectations 17
1.7 Heisenberg Uncertainty Principle 18
1.8 Observable and its Time Evolution 19
1.8.1 The Ehrenfest Theorem 21
1.9 Charge Density and Interatomic Potentials: The Hellmann–Feynman Theorem 22
1.10 The Pauli Exclusion Principle 24
1.11 On the Numerical Solution of the Schrödinger Equation 25
References 27

2 Ab-Initio Calculations 29
2.1 The Hartree–Fock Description 29
2.1.1 The Orbital Model 29
2.1.2 Accounting for the Pauli Exclusion Principle 30
2.1.3 The Variational Principle 31
2.1.4 A Direct Solution Procedure 32
2.1.5 The Hartree–Fock Approach 33
2.1.6 Post-Hartree–Fock Methods 34
2.2 Density Functional Theory 35
2.2.1 The First Hohenberg and Kohn Theorem 35
2.2.2 The Second Hohenberg and Kohn Theorem 37

2.2.3 The Hohn–Sham Equations ... 37
2.3 Concluding Remarks on the Quantum Scale ... 39
References ... 39

3 Coarse-Grained Descriptions ... 41
3.1 Molecular Dynamics ... 41
3.1.1 Some Simple Examples of Pair-Wise Interatomic Potentials ... 42
3.1.2 Integration Procedure ... 43
3.1.3 Discussion ... 44
3.1.4 Recovering Macroscopic Behaviors ... 46
3.1.5 Molecular Dynamics-Continuum Mechanics Bridging Techniques ... 51
3.1.6 Coarse-Grained Molecular Dynamics: DPD and MPCD ... 52
3.2 Brownian Dynamics: A Step Towards Coarse-Grained Models ... 55
3.2.1 The Langevin Equation ... 56
3.2.2 From Diffusion to Anomalous Diffusion ... 58
References ... 67

4 Kinetic Theory Models ... 69
4.1 Motivation ... 69
4.2 Kinetic Theory Description of Simple Liquids and Gases ... 70
4.2.1 Hydrodynamic Equations ... 72
4.2.2 The Lattice Boltzmann Method – LBM ... 75
4.3 Kinetic Theory Description of Some Complex Fluids ... 77
4.3.1 Dilute Suspensions of Non-Brownian Rods ... 78
4.3.2 On the Solution of the Fokker–Planck Equation ... 85
4.3.3 Dilute Suspensions of Brownian Rods ... 88
4.3.4 Rigid Clusters Composed of Rigid Rods ... 96
4.3.5 Extensible Rods ... 100
4.3.6 Dilute Polymers Solutions ... 103
4.3.7 Polymer Melts ... 109
4.3.8 Liquid Crystalline Polymers ... 111
4.3.9 Carbon-Nanotube Suspensions: Introducing Aggregation Effects ... 113
4.3.10 Kinetic Theory Approach to the Micro Structural Theory of Passive Mixing ... 115
4.4 The Chemical Master Equation ... 119
4.4.1 Moment Based Descriptions ... 121
4.4.2 From the Chemical Master Equation to Moment-Based Descriptions ... 122
References ... 124

Chapter 1
The Schrödinger Equation

Abstract The finest scale in the matter description consists of the one of quantum mechanics. In this chapter we revisit some valuable concepts of quantum mechanics, and more particularly the Schrödinger equation governing the time evolution of the so-called wavefunction, from which expectations can be easily derived. Another important output is the interatomic potentials responsible of the chemical bonds determining the structure and properties of materials. Even if the quantum framework is able to define the big picture, an important difficulty remains: the Schrödinger equation is defined in a highly multidimensional space and its solution is in most cases unattainable.

Keywords Wavefunction · Schrödinger equation · Expectations · Pauli principle Interatomic potentials

1.1 The History of Quantum Mechanics in Small Bits

Niels Bohr, one of the fathers of quantum mechanics, claimed that "if quantum mechanics doesn't profoundly shock you, you haven't understood it yet", and at the same time, he confessed that he himself never understood it completely. There are numerous books devoted to the history and central concepts of quantum mechanics [1–8]. In such a history, it is very difficult to define the starting point. In the particular case of the history of the atom, going back to the ideas of Democritus (four centuries before Christ) is, perhaps, a bit excessive!

The first relevant concepts were the Newtonian ideas about light in which he supposed it to be composed of particles. However, in the same period, Huygens claimed that light is, in fact, a wave. This last theory was widely accepted following the experiments of Thomas Young at the beginning of the 19th century and Augustin Fresnel a little bit later. This theory was definitively consolidated by Maxwell's works on electromagnetism.

The idea of the atom was first addressed by Robert Boyle in the 17th century, by Newton in his works on physics and optics, by Lavoisier in the 18th century,

© The Author(s) 2018

F. Chinesta and E. Abisset-Chavanne, *A Journey Around the Different Scales Involved in the Description of Matter and Complex Systems*, SpringerBriefs in Applied Sciences and Technology, https://doi.org/10.1007/978-3-319-70001-4_1

in his studies on combustion, by Dalton at the beginning of the 19th century, and by Gay-Lussac and Avogadro later on in the 19th century.

Atoms were considered by Maxwell and Boltzmann, in their theory of gases, to be the basis of statistical mechanics in physics. But these ideas were not accepted by the scientific community, with some recognized scientists going so far as to attack the theory, wich had not been confirmed experimentally at the time, where issue of reversibility was the major protagonist, motivating, apparently, Boltzmann suicide in 1906. The reality of atoms was established in one of the three capital works published by Einstein in 1905. The first of them concerned the theory of relativity, the second one the photoelectric effect (for which he received the Nobel Prize in 1921) and the third one the description of Brownian motion based on the existence of atoms in continuous movement.

At the end of the 19th century, a certain controversy arose about the essence of the radiation produced by a metallic filament driving an electrical current in a vacuum environment. J.J. Thompson of the Cavendish laboratory (founded by Maxwell in Cambridge, U.K.) conducted some key experiments that led to the discovery of the electron, which was the first elementary particle to be identified. For this discovery, J.J. Thompson received the Nobel Prize. Thompson, in the purest British tradition, imagined the atom as being much like a pudding, or even better, like a watermelon, in which the negative electrons were uniformly distributed throughout a positive matrix. Research was very active at that time on certain topics related to the nature of X-rays and radioactivity, with some of the day's most salient scientists becoming involved in it: Henry Becquerel, Marie and Pierre Curie and Ernest Rutherford. The last identified alpha and beta radiations (the gamma variety was discovered later) composed of positive and negative particles moving at high speed. Gamma rays seemed similar to X-rays but showed lower wavelengths.

At the beginning of the 20th century, Rutherford, while working at the University of Manchester, conducted some key experiments that provided evidence that the alpha beam scattered when it impacted gold foils. To explain the observed results, incompatible with the Thompson atom model, Rutherford proposed a model for the atom consisting of electrons rotating around a densely positive nucleus, and, in turn, he received the Nobel Prize. However, the Rutherford model had an important weakness, because an accelerated charged particle emits radiation. The related energy loss should motivate its collapse, in contrast with the experimental evidence on the atom's stability.

At that time, another important research topic concerned the radiation coming from a black body. Classical models predicted an infinite energy. The main contribution in this field was provided by Planck, who, using the statistical theory of thermodynamics developed by Boltzmann, concluded that the energy is quantified and related to the frequency ν from the expression $E = \mathrm{h}\nu$, where h is known as the Planck constant and its value in the metric system of units is $\mathrm{h} = 6.626 \times 10^{-34} (\mathrm{J} \cdot \mathrm{s})$.

Planck's works were considered in the model elaborated by Einstein (that earned him his Nobel Prize) concerning the photoelectric effect, which was explained by the introduction of the photon (light quanta). Thus, some centuries later, Newton's idea about the discrete essence of light (being composed of particles) was renewed.

A new proposal for the atom model was given by Niels Bohr, who suggested that electrons are located in orbitals that have a permitted energy according to energy quantification. To move from one orbit to another, it must receive or lose an amount of energy corresponding to an integer number of quanta. This simple model allowed for the successful description of the chemical structure of matter.

However, if we accept that light has a double character, particles and waves, why not extend this assertion to any other particle? This was the idea of Louis de Broglie, who, starting from Planck and Einstein's results, established in his PhD thesis that momentum, p, and frequency, ν, are related from $pc = \mathrm{h}\nu$, where c represents the speed of light. This relation has deep consequences, because momentum (characteristic of particles) appears on its left-hand side, while the right-hand side involves frequency (characteristic of waves), stating a subtle duality between particles and waves.

Obviously experimental corroboration of this was not simple, because in order to produce wave diffraction, one needs to proceed with a diaphragm whose dimension must be of the same order as that of the wavelength λ. From the de Broglie model, it results that $\lambda p = \mathrm{h}$. Thus, in order to obtain large enough wavelengths, momentum p must be small enough, as it is the case when considering lightweight particles. De Broglie, during his PhD defense, answering a question from a skeptical member of his evaluation committee, suggested that his hypothesis could be verified by diffracting electrons in a crystal, a scenario that makes the wavelength (atomic distance) and the particles' momentum compatible.

The crucial experiment suggested by de Broglie during his PhD dissertation was finally conducted simultaneously by Davisson in the USA and George Thompson (J.J. Thompson's son) in the UK. Both of them received the Nobel Prize. Thus, ironically, J.J. Thompson received the Nobel Prize for proving that electrons are particles, and his son, George Thompson, received the same prize, but for proving that electrons are "also" waves!

It is important to note that interference, observed in these experiments, and easily understood within the wave framework, can also be explained by the Feynman path-integral, in which each particle "simultaneously" follows any possible trajectory while interacting with itself.

Another eminent and key scientist was W. Pauli, who introduced the fourth quantum number (the one related to the spin), as well as the exclusion principle, which states that it is impossible to find two electrons in the same quantum state, allowing for a better understanding of Bohr's atom model. At that time, two statistics were introduced, the one by Bose–Einstein (that applies to particles with integer spin that are not concerned with the exclusion principle) and the one resulting from Fermi–Dirac (that applies to particles with semi-integer spin and for which the exclusion principle applies). These statistics are of capital importance for describing the structure of matter, as well as for explaining such exotic behaviors as the ones related to superconductivity and superfluidity.

The end of our brief overview concerns three attempts to describe the reality of quantum mechanics. The first approach was introduced in Gottingen by the team led by Max Born, who, in collaboration with Heisenberg, proposed the matrix mechanics to describe quantum dynamics. The second approach was initiated in Cambridge by Paul Dirac who, starting from Heisenberg's works, proposed a new algebra (quantum algebra) that incorporates matrix mechanics as a particular case. Finally, Erwin Schrödinger, inspired by de Broglie's works, proposed the introduction of a wavefunction describing the distribution of the particles in the whole space, as well as the partial differential equation governing its evolution, known as wave-equation. The last involves a continuous and unbounded medium (only the energy is discrete), in which the evolution of the scalar and complex unknown field (the wavefunction) is governed by a partial differential equation, so simple at first glance! Even if this last approach can be considered within the general formalism of quantum algebra, it seems conceptually simpler and more natural because of its connections with the well-established physics of waves.

Obviously the three conceptual schemes, the first two more particle-oriented and the last one clearly wave-oriented, are equivalent, as proved by Paul Dirac. However, the Schrödinger formalism continues to be more popular today, being preferred by a number of scientists working in the field of solid state physics and computational chemistry. Thus, this approach constitutes a powerful tool for describing the structure and mechanics of matter.

The wave-particle duality, the Heisenberg uncertainty principle (stating that a particle cannot simultaneously "have" a position and a momentum) and the reduction of the wavefunction (typically during an observation that produces the "instantaneous" reduction of the wavefunction in favor of an apparent material-particle), among many others, are at the origin of a number of paradoxes (some of them still open) that continue to baffle both passionate scientists and non- scientists! The interested reader can seek out some of the numerous available books previously referred to.

1.2 Planck Versus the Ultraviolet Catastrophe

In this section, we revisit the classical theory of black-body radiation (the Rayleigh–Jeans theory) and the so-called ultraviolet catastrophe. In order to solve this major issue, Planck introduced energy quantization, considered later by Einstein to explain the photoelectric effect.

1.2.1 Rayleigh–Jeans Theory

A stationary electric field in 1D writes

$$E = E_0 \sin\left(2\pi \frac{x}{\lambda}\right) \sin\left(2\pi \nu t\right), \tag{1.1}$$

where E_0 is the amplitude and λ and ν the wavelength and the frequency, respectively with $c = \lambda\nu$ (c being the speed of light). If the stationary wave is constrained to exist in an 1D cavity of length L with perfectly conductive boundaries, implying $E(x = 0) = E(x = L) = 0$, the following relation must be verified:

$$\frac{2L}{\lambda} = n, \quad n = 1, 2, 3, \ldots, \tag{1.2}$$

or

$$\nu = \frac{cn}{2L}, \quad n = 1, 2, 3, \ldots. \tag{1.3}$$

Thus, the number of allowed frequencies within the interval $[\nu, \nu + d\nu]$ becomes $\frac{2L}{c} d\nu$, in fact, the double $\frac{4L}{c} d\nu$, because we have two different polarizations.

When considering a cubic cavity $[0, L]^3$, and considering for a given direction expressed by the angles (α, β, γ) with respect to the cartesian axes

$$\begin{cases} \lambda = \lambda_x \cos\alpha \\ \lambda = \lambda_y \cos\beta \\ \lambda = \lambda_z \cos\gamma \end{cases}, \tag{1.4}$$

the compatibility conditions read as

$$\begin{cases} \frac{2L}{\lambda_x} = n_x \\ \frac{2L}{\lambda_y} = n_y \\ \frac{2L}{\lambda_z} = n_z \end{cases}, \tag{1.5}$$

or

$$\begin{cases} \frac{2L}{\lambda} \cos\alpha = n_x \\ \frac{2L}{\lambda} \cos\beta = n_y \\ \frac{2L}{\lambda} \cos\gamma = n_z \end{cases}. \tag{1.6}$$

Summing the square of all them (with $\cos^2\alpha + \cos^2\beta + \cos^2\gamma = 1$) results in

$$\left(\frac{2L}{\lambda}\right)^2 = n_x^2 + n_y^2 + n_z^2 \rightarrow \frac{2L}{\lambda} = \sqrt{n_x^2 + n_y^2 + n_z^2}, \tag{1.7}$$

or, with $r = \sqrt{n_x^2 + n_y^2 + n_z^2}$, in

$$r = \frac{2L}{c}\nu. \tag{1.8}$$

With the volume related to r being the one with the spherical shell, $4\pi r^2 dr$ divided by 8 for considering a single octant, and using the relation between r and the frequency ν, it turns out that the number of allowed frequencies in $[\nu, \nu + d\nu]$ is

$$\frac{\pi}{2}\left(\frac{2L}{c}\right)^3 \nu^2 d\nu, \tag{1.9}$$

in fact, the double, because of the two possible polarizations.

The energy density $\rho(\nu)$ (the total energy divided by the cavity volume L^3) can be computed using the energy equipartition theorem, which associates a kinetic energy (E_c) of $K_b T/2$ per degree of freedom:

$$\rho(\nu)d\nu = \frac{8\pi\nu^2 K_b T}{c^3} d\nu, \tag{1.10}$$

whose integral

$$\int_0^\infty \rho(\nu)d\nu = \int_0^\infty \frac{8\pi\nu^2 K_b T}{c^3} d\nu \tag{1.11}$$

diverges, i.e.,

$$\int_0^\infty \rho(\nu)d\nu = \infty, \tag{1.12}$$

which constitutes the so-called ultraviolet catastrophe.

1.2.2 *Planck Theory*

To avoid the divergence of the integral, Planck proposed a discrete energy distribution

$$E(\nu) = nh\nu, \quad n = 0, 1, 2, 3, \ldots, \tag{1.13}$$

which, using the Boltzmann theory, provides at equilibrium the following probability distribution:

$$p(n) = \frac{e^{-\frac{E_n}{K_b T}}}{\sum_{i=0}^{\infty} e^{-\frac{E_i}{K_b T}}}, \tag{1.14}$$

which is the probability, at equilibrium, of having n photons related to the frequency ν.

Now, the energy average $\overline{E}_\nu$ read as

$$\overline{E}_\nu = \sum_{n=0}^{\infty} E_n p(n) = \frac{\sum\limits_{n=0}^{\infty} nh\nu e^{-\frac{E_n}{K_b T}}}{\sum\limits_{n=0}^{\infty} e^{-\frac{E_n}{K_b T}}}, \tag{1.15}$$

which using the change of variable $x = e^{-\frac{h\nu}{K_b T}}$, becomes

$$\overline{E}_\nu = h\nu \frac{\sum\limits_{n=0}^{\infty} nx^n}{\sum\limits_{n=0}^{\infty} x^n} = h\nu \frac{x + 2x^2 + 3x^3 + \cdots}{1 + x + x^2 + \cdots} = h\nu x \frac{1 + 2x + 3x^2 + \cdots}{1 + x + x^2 + \cdots}. \tag{1.16}$$

Taking into account the relations

$$\begin{cases} \frac{1}{1-x} = 1 + x + x^2 + \cdots \\ \frac{1}{(1-x)^2} = 1 + 2x + 3x^2 + \cdots \end{cases}, \tag{1.17}$$

the energy average read as

$$\overline{E}_\nu = \frac{h\nu x}{1-x} = \frac{h\nu}{e^{\frac{h\nu}{K_b T}} - 1}. \tag{1.18}$$

The energy density then read as

$$\rho(\nu)d\nu = \frac{8\pi\nu^2 \overline{E}_\nu}{c^3} d\nu = \frac{8\pi\nu^2}{c^3} \frac{h\nu}{e^{\frac{h\nu}{K_b T}} - 1} d\nu, \tag{1.19}$$

whose integral does not diverge anymore:

$$\rho = \int_0^\infty \rho(\nu)d\nu = \frac{8\pi^5}{15h} \left(\frac{K_b T}{h}\right)^4, \tag{1.20}$$

in which the Stefan–Boltzmann law $\rho \propto T^4$ can be seen.

1.3 An Intuitive Approach to the Schrödinger Equation

Even if the Schrödinger equation could be assumed to be a first principe (avoiding the necessity of deriving it), in this section we prefer to derive it by following a non-rigorous procedure, which nonetheless has the merit of being almost intuitive.

We consider as our starting point some key results:

- The de Broglie equation establishing the relation between momentum p and wavelength λ: $\lambda p = h$;
- The Einstein equation establishing the relation between energy and frequency (according to Planck): $E = h\nu$;
- The general expression of a sinusoidal traveling wave:

$$\Psi(x, t) = \sin 2\pi \left(\frac{x}{\lambda} - \nu t\right); \tag{1.21}$$

- The non-relativistic expression of energy:

$$E = \frac{p^2}{2m} + V, \tag{1.22}$$

with V as the potential. Because of the fact that we are using the non-relativistic expression of the energy, the resulting Schrödinger equation becomes non-relativistic. Dirac considered the energy expression's relativistic counterpart and derived the relativistic counterpart of the Schrödinger equation (the so-called Dirac equation).

Now, we assume a linear physics and focus on the simplest scenario concerning a particle evolving in the space with constant potential V_0. In the absence of interactions (forces resulting from the potential gradient, here vanishing), the momentum and the energy should remain constant, according to the de Broglie and Einstein expressions $p\lambda = h$ and $E = h\nu$. Thus, the energy expression (1.22) becomes

$$\frac{h^2}{2m\lambda^2} + V_0 = \nu h, \tag{1.23}$$

or by introducing $k = \frac{2\pi}{\lambda}$, $\omega = 2\pi\nu$ and $\hbar = \frac{h}{2\pi}$, the energy expression (1.23) can be rewritten as

$$\frac{\hbar^2 k^2}{2m} + V_0 = \omega\hbar, \tag{1.24}$$

and the sinusoidal wave (1.21) read as

$$\Psi(x, t) = \sin(kx - \omega t). \tag{1.25}$$

Now, in absence of interactions, the evolution of the wavefunction $\Psi(x, t)$ is expected to be given by a travelling wave like (1.25). Its space and time derivatives read as

$$\begin{cases} \frac{\partial \Psi}{\partial x} = k\cos(kx - \omega t) \\ \frac{\partial^2 \Psi}{\partial x^2} = -k^2 \sin(kx - \omega t) \\ \frac{\partial \Psi}{\partial t} = -\omega\cos(kx - \omega t) \end{cases}, \tag{1.26}$$

whose comparison with the energy expression (1.24) suggests

$$\alpha \frac{\partial^2 \Psi}{\partial x^2} + V\Psi = \beta \frac{\partial \Psi}{\partial t}. \tag{1.27}$$

To compute the expression of α and β, the wavefunction (1.25) is substituted into (1.27) with $V = V_0$, leading to

$$-\alpha k^2 \sin(kx - \omega t) + \sin(kx - \omega t) V_0 = -\beta \omega \cos(kx - \omega t), \tag{1.28}$$

which has no solution for any choice of α and β.

To increase our chances, we consider a more general form of the traveling wave

$$\Psi(x, t) = \cos(kx - \omega t) + \gamma \sin(kx - \omega t). \tag{1.29}$$

Introducing this expression into (1.27) yields

$$(-\alpha k^2 + V_0 + \beta\omega\gamma)\cos(kx - \omega t) + (-\alpha k^2 \gamma + V_0 \gamma - \beta\omega)\sin(kx - \omega t) = 0, \tag{1.30}$$

which implies

$$\begin{cases} -\alpha k^2 + V_0 = -\beta\omega\gamma \\ -\alpha k^2 + V_0 = \frac{\beta\omega}{\gamma} \end{cases}. \tag{1.31}$$

Subtracting both Eq. (1.31) results in

$$-\gamma - \frac{1}{\gamma} = 0, \tag{1.32}$$

which implies

$$\gamma = -\frac{1}{\gamma}, \tag{1.33}$$

that is, $\gamma^2 = -1$, or $\gamma = \pm i$. Substituting this value in the first equation in (1.31) results in

$$-\alpha k^2 + V_0 = \pm i\beta\omega, \tag{1.34}$$

whose comparison with the expression of the energy (1.24) leads to

$$\begin{cases} \alpha = -\frac{\hbar^2}{2m} \\ \beta = \pm i\hbar \end{cases}. \tag{1.35}$$

Without loss of generality, we choose β to be positive. We can now write the Schrödinger equation:

$$-\frac{\hbar^2}{2m}\frac{\partial^2\Psi}{\partial x^2} + V\Psi = i\hbar\frac{\partial\Psi}{\partial t}. \tag{1.36}$$

It is important to note that as the Heisenberg uncertainty principle, discussed in the next chapter, establishes a huge difference between classical and quantum particles (the latter cannot have a defined position and momentum simultaneously), in the wave description just described, the wavefunction seems also to be quite a strange wave, with a part of it defined in the complex dimension.

1.4 The Feynman Approach

Feynman considered that for a free particle, the wavefunction evolves from its initial state $\Psi(y, t = 0)$ to the final one $\Psi(x, t)$:

$$\Psi(y, t = 0) \rightarrow \Psi(x, t). \tag{1.37}$$

The assumption of linearity implies that, given two evolutions,

$$\begin{cases} \Psi_I(y, t = 0) \rightarrow \Psi_I(x, t) \\ \Psi_{II}(y, t = 0) \rightarrow \Psi_{II}(x, t) \end{cases}, \tag{1.38}$$

it results that

$$\alpha\Psi_I(y, t = 0) + \beta\Psi_{II}(y, t = 0) \rightarrow \alpha\Psi_I(x, t) + \beta\Psi_{II}(x, t). \tag{1.39}$$

From the Schwartz kernel theorem, we can write

$$\Psi(x, t) = \int_y \mathcal{G}(x, y; t)\Psi(y, t = 0)dy, \tag{1.40}$$

with the kernel function $\mathcal{G}(\bullet)$ independent of the origin of the coordinate system, i.e., $\mathcal{G}(x, y; t) = \mathcal{G}(x - y; t)$, and $\mathcal{G}(x, y; t) = \mathcal{G}\left((x - y)^2; t\right)$ in the case of assuming isotropy. A dimensional analysis allows us to write

$$\mathcal{G}(x, y; t) = \mathcal{G}\left(\frac{(x - y)^2 m}{\hbar t}\right). \tag{1.41}$$

Now, we consider a final state attained from two different paths,

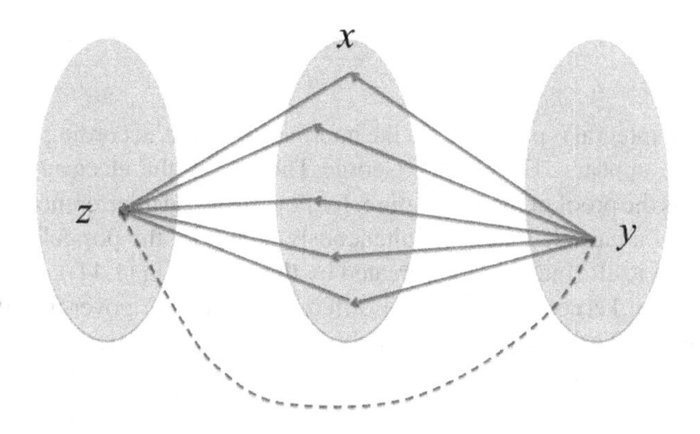

Fig. 1.1 Evolution from $\Psi(y; t = 0)$ to $\Psi(z; t + t')$

$$\Psi(z; t + t') = \int_y \mathcal{G}(z, y; t + t')\Psi(y, t = 0)dy, \tag{1.42}$$

and

$$\Psi(z; t + t') = \int_x \mathcal{G}(z, x; t')\Psi(x, t)dx, \tag{1.43}$$

which imply the composition rule

$$\mathcal{G}(z, y; t + t') = \int_x \mathcal{G}(z, x; t')\mathcal{G}(x, y; t)dx, \tag{1.44}$$

illustrated in Fig. 1.1.

Using the previous results (1.41) and (1.44) with the normality condition

$$\int_x \overline{\Psi}(x, t)\Psi(x, t)dx = 1, \tag{1.45}$$

with $\overline{\Psi}$ the conjugate of Ψ, the expression of the kernel can be derived:

$$\mathcal{G}(x, y; t) = \frac{i}{t}\frac{m}{2\pi\hbar}e^{\frac{i}{t}\frac{(x-y)^2 m}{\hbar}}, \tag{1.46}$$

where reversibility is ensured because $\frac{i}{t} = \frac{-i}{-t}$.

Thus, if the path $y \to x$ implies a phase change

$$\varphi = \frac{(x - y)^2 m}{\hbar t}, \tag{1.47}$$

$x \to z$ increases the phase change

$$\varphi' = \varphi + \frac{(z-x)^2 m}{\hbar t'}, \tag{1.48}$$

but because the integral is performed at the intermediate state, according to Eq. (1.44), an interference appears. Thus, if for George Thompson, the electron interference experience was the proof of its wave nature, for Feynman, the interference was simply the signature of a particle that is simultaneously taking all the possible trajectories interacting with itself (interaction is created by the integral in (1.44)).

Now, from the kernel identified, the differential equation governing the evolution of the wavefunction can be obtained, and as expected, it coincides with the Schrödinger equation.

1.5 The Schrödinger Equation

In classical mechanics, the evolution of a particle can be determined from the knowledge of its position $\mathbf{r}$ and velocity $\mathbf{v}$ (or its momentum $\mathbf{p} = m\mathbf{v}$, with m the particle mass) at a point of its trajectory. However, in quantum mechanics, position and momentum can not be simultaneously associated with a particle, according to the Heisenberg uncertainty principle. Thus, rather than speaking about position or momentum, we should speak about probability distribution related to position $P_r = |\Psi|^2$ and momentum $P_p = |\Psi_p|^2$. The necessity of considering the square of Ψ or Ψ_p results from the fact that both functions are complex, and the resulting probabilities P_r and P_p must necessarily be real numbers.

Precisely because of the uncertainty principle, these two wavefunctions Ψ and Ψ_p are not independent, the first one being the Fourier transform of the second one. The properties of the Fourier transform guarantee that the higher the localization in space, the lower its localization in the momentum variable and vice-versa [9]. Of course, from the physical point of view, this fact induces a conceptual difficulty, because in classical mechanics, two pieces of information are needed to determine the evolution of the system (i.e., position and velocity) and the quantum system seems to require only one Ψ (or Ψ_p, because both are related to the Fourier transform). This paradox is only apparent because both wavefunctions are complex, and thus they have both a real part and an imaginary part. Thus, knowledge of a quantum system requires knowing the evolution of the wavefunction Ψ (or Ψ_p) introduced by Erwin Schrödinger in the 1920s, which only needs the solution of the equation governing its distribution or its evolution in the transient case.

For the sake of simplicity, we will first introduce the Schrödinger equation considering neither the relativistic effects nor the spin. If we assume a system composed of N_P particles, the evolution of the joint wavefunction $\Psi = \Psi\left(\mathbf{r}_1, \mathbf{r}_2, \ldots, \mathbf{r}_{N_P}, t\right)$ is governed by the Schrödinger equation whose dimensionless form results [10]:

$$i\frac{\partial \Psi}{\partial t} = -\sum_{p=1}^{N_p} \frac{1}{2m_p} \nabla_p^2 \Psi + \sum_{p=1}^{N_p} \sum_{k=p+1}^{N_p} V_{pk} \Psi, \tag{1.49}$$

where each particle p is distributed in the whole physical space $\mathbf{r}_p = (x_p, y_p, z_p) \in \mathbb{R}^3$ and $i = \sqrt{-1}$. The differential operator ∇_p^2 is defined in the conformational space of each particle, i.e.,

$$\nabla_p^2 = \frac{\partial^2}{\partial x_p^2} + \frac{\partial^2}{\partial y_p^2} + \frac{\partial^2}{\partial z_p^2}. \tag{1.50}$$

The Coulomb potential describing the inter-particle interactions writes

$$V_{pk} = \frac{q_p \cdot q_k}{||\mathbf{r}_p - \mathbf{r}_k||}, \tag{1.51}$$

where the masses m_p are unity for electrons, charges q_j are -1 for electrons and $+Z_j$ (atomic numbers) for nuclei.

Equation (1.49) can be rewritten in the compact form

$$(\mathscr{H} - \mathscr{T})\,\Psi = 0, \tag{1.52}$$

where $\mathscr{H}$ denotes the Hamiltonian operator and $\mathscr{T} = -i\frac{\partial}{\partial t}$. This equation was proposed, not deduced, and it is today considered a first principle, like the Newton equation in classical mechanics.

It is important to note that Eq. (1.49) can be integrated from an initial condition $\Psi(\mathbf{x}, t = 0)$, leading to $\Psi(\mathbf{x}, t)$, with $t \in (0, T]$. If we consider $0 < \tau < T$, and we do not observe the state of the system at $t = \tau$, then for $t > \tau$, $\Psi(\mathbf{x}, t)$ corresponds to the solution of Eq. (1.52) with the initial condition $\Psi(\mathbf{x}, t = 0)$. However, if we observe the system at time $t = \tau$, the observation process irreversibly modifies the wavefunction, that is, $\Psi(\mathbf{x}, \tau - \varepsilon) \neq \Psi(\mathbf{x}, \tau + \varepsilon)$ (with $\varepsilon \to 0$). Then, after the observation, the system is described by the solution to Eq. (1.52) with the initial condition $\Psi(\mathbf{x}, \tau + \varepsilon)$ that differs from the solution $\Psi(\mathbf{x}, t)$ computed from $\Psi(\mathbf{x}, t = 0)$.

By assuming the separation of variables, the wavefunction can be decomposed according to

$$\Psi = \Psi_r\left(\mathbf{r}_1, \ldots, \mathbf{r}_{N_p}\right) \cdot \Psi_t(t). \tag{1.53}$$

Introducing this expression into the Schrödinger equation (1.52) and dividing by $\Psi_r \cdot \Psi_t$ yields

$$\frac{\Psi_t \cdot \mathscr{H}(\Psi_r)}{\Psi_r \cdot \Psi_t} - \frac{\Psi_r \cdot \mathscr{T}(\Psi_t)}{\Psi_r \cdot \Psi_t} = 0, \tag{1.54}$$

or

$$\frac{\mathscr{H}(\Psi_{\mathbf{r}})}{\Psi_r} - \frac{\mathscr{T}(\Psi_t)}{\Psi_t} = 0. \tag{1.55}$$

The first term depends on the space variables and the second one on the time, implying that both terms are equal to a constant E (which, as will be discussed later, represents the energy), that is,

$$\begin{cases} \frac{\mathscr{T}(\Psi_t)}{\Psi_t} = E \\ \\ \frac{\mathscr{H}(\Psi_r)}{\Psi_r} = E \end{cases}. \tag{1.56}$$

The integration of the first relation in (1.56) implies

$$\Psi_t(t) = A\, \mathrm{e}^{-iEt}, \tag{1.57}$$

and the second one leads to the following eigenproblem:

$$\mathscr{H}(\Psi_r(\mathbf{r}_1, \ldots, \mathbf{r}_{Np})) = E\Psi_r(\mathbf{r}_1, \ldots, \mathbf{r}_{Np}). \tag{1.58}$$

The eigenfunctions $\Psi_r^{(n)}(\mathbf{r}_1, \ldots, \mathbf{r}_{Np})$ related to the eigenvalues E_n define an orthogonal basis. Obviously, each function

$$\Psi^{(n)} = A_n\, e^{-iE_n t} \cdot \Psi_r^{(n)}(\mathbf{r}_1, \ldots, \mathbf{r}_{Np}), \tag{1.59}$$

is time-independent, because the physical meaning of the solution is done by the complex norm of the wavefunction, that is,

$$|\Psi^{(n)}| = \overline{\Psi^{(n)}} \cdot \Psi^{(n)} = A_n^2\, e^{-iE_n t} \cdot e^{iE_n t} \cdot \overline{\Psi_r^{(n)}} \cdot \Psi_r^{(n)} = A_n^2\, \overline{\Psi_r^{(n)}} \cdot \Psi_r^{(n)}, \tag{1.60}$$

where $\overline{a}$ denotes the conjugate of a.

Of course, the general transient solution can be written as

$$\Psi(\mathbf{r}_1, \ldots, \mathbf{r}_{N_p}, t) = \sum_{n=1}^{\infty} A_n\, e^{-iE_n t} \cdot \Psi_r^{(n)}(\mathbf{r}_1, \ldots, \mathbf{r}_{N_p}), \tag{1.61}$$

which is obviously not time-independent anymore.

In what follows, we assume a system composed of N_e electrons and N_n nuclei, and a time-independent solution of the Schrödinger equation. The Born–Oppenheimer model writes the total wavefunction of the combined system (electrons and nuclei) as a sum of products of wavefunctions related to the electrons and to the nuclei. Born and Oppenheimer showed that the total wavefunction can be reasonably approximated from a single product of a nuclei wavefunction $\Psi_n = \Psi_n(\mathbf{R}_1, \ldots, \mathbf{R}_{N_n})$ and an electronic wavefunction $\Psi_e = \Psi_e(\mathbf{r}_1, \ldots, \mathbf{r}_{N_e}; \mathbf{R}_1, \ldots, \mathbf{R}_{N_n})$ that depends parametrically on the nuclei coordinates. Introducing $\Psi = \Psi_n \cdot \Psi_e$ into the Schrödinger's eigenproblem $\mathscr{H}(\Psi) = E\Psi$ results in

$$\Psi_n \cdot \mathcal{H}_e(\Psi_e) + \Psi_e \cdot \mathcal{H}_n(\Psi_n) = E\, \Psi_e \cdot \Psi_n, \tag{1.62}$$

or

$$\frac{\mathcal{H}_e(\Psi_e)}{\Psi_e} + \frac{\mathcal{H}_n(\Psi_n)}{\Psi_n} = E, \tag{1.63}$$

where the first term depends on the electron coordinates and the second one on the nuclei coordinates. Thus, we can write

$$\frac{\mathcal{H}_e(\Psi_e)}{\Psi_e} = E_e, \tag{1.64}$$

implying

$$\mathcal{H}_e(\Psi_e) = E_e\, \Psi_e, \tag{1.65}$$

and

$$\frac{\mathcal{H}_n(\Psi_n)}{\Psi_n} = E - E_e, \tag{1.66}$$

which results in

$$(\mathcal{H}_n + E_e)\Psi_n = E\, \Psi_n. \tag{1.67}$$

The difference between the dynamics of electrons and nuclei is derived in the following way: (i) the nuclei are typically tens of thousands times heavier than the electrons; (ii) the particles constituting the molecules are in equilibrium, so, on average, they have similar kinetic energies (equipartition theorem); (iii) thus, the ratio of the square of their velocities (electrons and nuclei) will be roughly the inverse ratio of their masses (velocity ratio on the order of hundreds); (iv) then the wavelength associated with the nuclei is hundreds of times lower than the one corresponding to the electrons. The last conclusion comes from the fact that, according to de Broglie, $\lambda_n^{-1} \sim p_n = m_n v_n \approx 10^4 m_e 10^{-2} v_e = 10^2 p_e \sim 10^2 \lambda_e^{-1}$ (where $\bullet_n$ and $\bullet_e$ refer, respectively to the properties of the nuclei and electrons).

Thus, electrons see stationary nuclei and nuclei see electrons distributed (in a sort of mean). From now on, we assume the nuclei to be fixed in the physical space and write

$$\mathcal{H}(\Psi(\mathbf{r}_1, \ldots, \mathbf{r}_{N_e}; \mathbf{R}_1, \ldots, \mathbf{R}_{N_n})) = E\, \Psi(\mathbf{r}_1, \ldots, \mathbf{r}_{N_e}; \mathbf{R}_1, \ldots, \mathbf{R}_{N_n}), \tag{1.68}$$

where the parametrical dependence of the electronic distribution on the nuclei position has been emphasized. The Hamiltonian can be written as

$$\mathscr{H}(\Psi) = -\sum_{e=1}^{N_e} \frac{1}{2}\nabla_e^2 \Psi + \sum_{e=1}^{N_e}\sum_{n=1}^{N_n} V_{en}\Psi + \sum_{e=1}^{N_e}\sum_{e'=e+1}^{N_e} V_{ee'}\Psi, \quad (1.69)$$

where

$$V_{ee'} = \frac{1}{||\mathbf{r}_e - \mathbf{r}_{e'}||} \quad (1.70)$$

and

$$V_{en} = -\frac{Z}{||\mathbf{r}_e - \mathbf{R}_n||}. \quad (1.71)$$

1.6 Relations Between Position and Momentum Wavefunctions

As previously indicated, $\Psi(\mathbf{x}, t)$ and $\Psi_p(\mathbf{p}, t)$ are Fourier transforms of each other in order to fulfill the Heisenberg uncertainty principle addressed later. Thus, proceeding in the 1D case, for the sake of simplicity, we have

$$\Psi_p(p, t) = \frac{1}{\sqrt{2\pi\hbar}} \int_{-\infty}^{\infty} \Psi(x, t) e^{-i\frac{px}{\hbar}} dx = \mathscr{F}(\Psi(x, t)), \quad (1.72)$$

where $\mathscr{F}(\bullet)$ refers to the Fourier transform and the use of the constant $\hbar$ ($\hbar = \frac{h}{2\pi}$) is introduced into the Fourier transform definition for addressing the Heisenberg uncertainty principle later. The inverse transformation writes

$$\Psi(x, t) = \frac{1}{\sqrt{2\pi\hbar}} \int_{-\infty}^{\infty} \Psi_p(p, t) e^{i\frac{px}{\hbar}} dx. \quad (1.73)$$

This definition of the Fourier transform is not practical when using standard convolution expressions, but this is not an issue here, and moreover, the convolution expression can be modified accordingly.

Using the Fourier transform properties, we can write

$$p^n \Psi_p = \mathscr{F}\left(\left(\frac{\hbar}{i}\right)^n \frac{\partial^n \Psi}{\partial x^n}\right). \quad (1.74)$$

In what follows, we will also make use of Parseval's theorem that, for $\Psi_p = \mathscr{F}(\Psi)$ and $\Phi_p = \mathscr{F}(\Phi)$, states

$$\int_{-\infty}^{\infty} \overline{\Psi}_p \cdot \Phi_p \, dp = \int_{-\infty}^{\infty} \overline{\Psi} \cdot \Phi \, dx. \tag{1.75}$$

1.6.1 Calculating Expectations

As soon as the electronic distribution Ψ is known, one could compute the one related to the electronic momentum Ψ_p, and from both, the different expected values of position, momentum, energy, etc. Imagine a simple system consisting of a single electron in the one-dimensional space, i.e., $\mathbf{r} = x$. Thus, writing $P_x = |\Psi|^2$ and $P_p = |\Psi_p|^2$, the expected value for the position reads as

$$\langle x \rangle = \int_{\mathbb{R}} x P_x \, dx = \int_{\mathbb{R}} x \, \overline{\Psi}(x) \cdot \Psi(x) \, dx, \tag{1.76}$$

with $\mathbb{R} = (-\infty, \infty)$.

The analogue expression for the momentum results in

$$\langle p \rangle = \int_{\mathbb{R}} p P_p \, dx = \int_{\mathbb{R}} p \, \overline{\Psi}_p(p) \cdot \Psi_p(p) \, dp, \tag{1.77}$$

which using Eq. (1.74) and the Parseval theorem, results in

$$\langle p \rangle = \int_{\mathbb{R}} p P_p \, dp = \int_{\mathbb{R}} \overline{\Psi}(x) \frac{\hbar}{i} \frac{\partial \Psi(x)}{\partial x} \, dx. \tag{1.78}$$

Analogously, we can compute $\langle p^2 \rangle$:

$$\langle p^2 \rangle = \int_{\mathbb{R}} \overline{\Psi}(x) \left(\frac{\hbar}{i} \right)^2 \frac{\partial^2 \Psi(x)}{\partial x^2} \, dx. \tag{1.79}$$

Thus, for a polynomial $g(p)$, it results that

$$\langle g \rangle = \int_{\mathbb{R}} \overline{\Psi}(x) \, g \left(\frac{\hbar}{i} \frac{\partial}{\partial x} \right) \Psi(x) \, dx, \tag{1.80}$$

where $g\left(\frac{\hbar}{i}\frac{\partial}{\partial x}\right)$ refers to the replacement of p in the polynomial $g(p)$ by the differential operator $\frac{\hbar}{i}\frac{\partial}{\partial x}$.

By considering $f(x)$ and $g(p)$ and using the linearity, it results that

$$\langle f+g\rangle = \int_{\mathbb{R}} \overline{\Psi}(x) \left(f + g\left(\frac{\hbar}{i}\frac{\partial}{\partial x}\right)\right) \Psi(x)\, dx. \tag{1.81}$$

When applied to the Hamiltonian $\mathscr{H}(x, p)$

$$\mathscr{H} = \frac{p^2}{2m} + V(x), \tag{1.82}$$

it results that

$$\langle \mathscr{H}\rangle = \int_{\mathbb{R}} \overline{\Psi}(x)\mathscr{H}\Psi(x)\, dx = \int_{\mathbb{R}} \overline{\Psi}(x) \left(V(x) - \frac{\hbar^2}{2m}\frac{\partial^2}{\partial x^2}\right) \Psi(x)\, dx. \tag{1.83}$$

All the previous results can be easily extended to 3D scenarios. Thus, as soon as the wavefunction $\Psi(\mathbf{x}, t)$ is known, the different expectations related to position, momentum and/or energy can be easily computed.

1.7 Heisenberg Uncertainty Principle

If we assume for a while (and without loss of generality) $\langle x\rangle = 0$ and $\langle p\rangle = 0$,

$$(\Delta x)^2 = \langle (x - \langle x\rangle)^2\rangle = \langle x^2\rangle, \tag{1.84}$$

and similarly

$$(\Delta p)^2 = \langle p^2\rangle. \tag{1.85}$$

If we define the integral

$$\mathscr{I}(\alpha) = \int_{\mathbb{R}} \left| \left(\alpha\hbar\frac{\partial}{\partial x} - x\right)\Psi\right|^2 dx \geq 0, \tag{1.86}$$

its expansion reads as

$$\mathscr{I}(\alpha) = \int_{\mathbb{R}} \left(\alpha\hbar\frac{\partial}{\partial x} - x\right)\overline{\Psi} \cdot \left(\alpha\hbar\frac{\partial}{\partial x} - x\right)\Psi\, dx$$

$$= \alpha^2\hbar^2 \int_{\mathbb{R}} \frac{\partial\overline{\Psi}}{\partial x}\frac{\partial\Psi}{\partial x}\, dx + \int_{\mathbb{R}} x^2\overline{\Psi}\Psi\, dx - \alpha\hbar\left(\int_{\mathbb{R}} \frac{\partial\overline{\Psi}}{\partial x} x\Psi\, dx + \int_{\mathbb{R}} x\overline{\psi}\frac{\partial\Psi}{\partial x}\right). \tag{1.87}$$

Now, by integrating by parts the first term on the right-hand side and taking into account that the wavefunction vanishes when $\mathbf{x} \to \infty$, it results that

$$\hbar^2 \int_{\mathbb{R}} \frac{\partial \overline{\Psi}}{\partial x} \frac{\partial \Psi}{\partial x} \, dx = -\hbar^2 \int_{\mathbb{R}} \overline{\Psi} \frac{\partial^2 \Psi}{\partial x^2} \, dx = (\Delta p)^2. \tag{1.88}$$

On the other hand, integrating by parts the third term yields

$$\int_{\mathbb{R}} \frac{\partial \overline{\Psi}}{\partial x} x \Psi \, dx = -\int_{\mathbb{R}} \overline{\Psi} \left(\Psi + x \frac{\partial \Psi}{\partial x} \right) dx = -1 - \int_{\mathbb{R}} \overline{\Psi} x \frac{\partial \Psi}{\partial x} \, dx, \tag{1.89}$$

leading to the polynomial expression

$$\mathscr{I}(\alpha) = \alpha^2 (\Delta p)^2 + \alpha \hbar + (\Delta x)^2 \geq 0. \tag{1.90}$$

Since this quantity is nonnegative, the discriminant of the quadratic polynomial

$$\hbar^2 - 4(\Delta p)^2 (\Delta x)^2 \leq 0 \tag{1.91}$$

is negative, which leads to the usual expression of the Heisenberg uncertainty principle

$$(\Delta p)^2 (\Delta x)^2 \geq \frac{\hbar^2}{4}. \tag{1.92}$$

When $\langle x \rangle \neq 0$, and/or $\langle p \rangle \neq 0$, we obtain the same result as proved in [9]. All the previous results can be easily extended to 3D scenarios.

The main consequences are that quantum particles are quite different from their classical counterpart, because in quantum theory, position and velocity cannot exist simultaneously; if position is perfectly defined (it exists in a classical sense), the "quantum particle" has a totally undetermined velocity and vice-versa.

There exists another uncertainty relationship relating energy and time (even if the latter is not an operator). This expression gives an approximate value for the length of time taken for the energy to change by one standard deviation. In order to get an accurate energy measurement, it should change slowly. This result is of major interest for explaining the quantum tunnel effect.

1.8 Observable and its Time Evolution

The expected value of any observable F (Hermitian operator on a complex vector space, like position, energy, ...),

$$\langle F \rangle = \int_{\mathbb{R}} \overline{\Psi} F \Psi \, dx, \tag{1.93}$$

can be written in a more compact form as

$$\langle F \rangle = (\Psi, F\Psi), \tag{1.94}$$

where

$$(\Psi, \Phi) = \int_{\mathbb{R}} \overline{\Psi} \cdot \Phi \, dx. \tag{1.95}$$

When considering the time-independent Hamiltonian observable $\langle \mathscr{H} \rangle = (\Psi, \mathscr{H}\Psi)$, the time evolution of its expectation is given by

$$\frac{d\langle \mathscr{H} \rangle}{dt} = \left(\frac{\partial \Psi}{\partial t}, \mathscr{H}\Psi \right) + \left(\Psi, \mathscr{H}\frac{\partial \Psi}{\partial t} \right), \tag{1.96}$$

where the time-independent nature of $\mathscr{H}$ implies $\frac{\partial \mathscr{H}}{\partial t} = 0$.

By integrating by parts and taking into account both the Hermitian character of observables (the Hamiltonian in the present case) and the fact that Ψ vanishes when $x \to \infty$ gives us

$$\frac{d\langle \mathscr{H} \rangle}{dt} = \left(\frac{\partial \Psi}{\partial t}, \mathscr{H}\Psi \right) + \left(\mathscr{H}\Psi, \frac{\partial \Psi}{\partial t} \right), \tag{1.97}$$

which using the Schrödinger equation $\mathscr{H}\Psi = i\hbar \frac{\partial \Psi}{\partial t}$, leads to

$$\frac{d\langle \mathscr{H} \rangle}{dt} = \left(\frac{\partial \Psi}{\partial t}, i\hbar\frac{\partial \Psi}{\partial t} \right) + \left(i\hbar\frac{\partial \Psi}{\partial t}, \frac{\partial \Psi}{\partial t} \right) = 0, \tag{1.98}$$

which states the energy conservation.

For any time-independent observable $\mathscr{O}$ ($\mathscr{O}$ does not depend on t), it results that

$$\langle \mathscr{O} \rangle = \int_{\mathbb{R}} \overline{\Psi} \mathscr{O} \Psi \, dx, \tag{1.99}$$

or using the notation previously introduced,

$$\langle \mathscr{O} \rangle = (\Psi, \mathscr{O}\Psi). \tag{1.100}$$

The time evolution of observable $\mathscr{O}$ reads as

$$\frac{d\langle \mathscr{O} \rangle}{dt} = \left(\frac{\partial \Psi}{\partial t}, \mathscr{O}\Psi \right) + \left(\Psi, \mathscr{O}\frac{\partial \Psi}{\partial t} \right), \tag{1.101}$$

which, using the Schrödinger equation again, gives us

$$\frac{d\langle \mathscr{O} \rangle}{dt} = \left(-\frac{i}{\hbar}\mathscr{H}\Psi, \mathscr{O}\Psi \right) + \left(\Psi, -\frac{i}{\hbar}\mathscr{O}\mathscr{H}\Psi \right), \tag{1.102}$$

which can be rewritten as

$$\frac{d\langle\mathcal{O}\rangle}{dt} = \frac{i}{\hbar}\left((\mathcal{H}\Psi, \mathcal{O}\Psi) - (\Psi, \mathcal{O}\mathcal{H}\Psi)\right), \tag{1.103}$$

or

$$\frac{d\langle\mathcal{O}\rangle}{dt} = \frac{i}{\hbar}\left((\Psi, \mathcal{H}\mathcal{O}\Psi) - (\Psi, \mathcal{O}\mathcal{H}\Psi)\right), \tag{1.104}$$

which ultimately leads to

$$\frac{d\langle\mathcal{O}\rangle}{dt} = \frac{i}{\hbar}\left(\Psi, (\mathcal{H}\mathcal{O} - \mathcal{O}\mathcal{H})\,\Psi\right), \tag{1.105}$$

or its compact counterpart

$$\frac{d\langle\mathcal{O}\rangle}{dt} = \frac{i}{\hbar}\langle[\mathcal{H}, \mathcal{O}]\rangle, \tag{1.106}$$

where the so-called commutator $[\bullet]$ (Poisson bracket) is defined as

$$[\mathcal{H}, \mathcal{O}] = \mathcal{H}\mathcal{O} - \mathcal{O}\mathcal{H}. \tag{1.107}$$

When the observable depends explicitly on time, it results that

$$\frac{d\langle\mathcal{O}\rangle}{dt} = \left\langle\frac{d\mathcal{O}}{dt}\right\rangle + \frac{i}{\hbar}\langle[\mathcal{H}, \mathcal{O}]\rangle. \tag{1.108}$$

1.8.1 The Ehrenfest Theorem

From these results, it is very easy to prove Ehrenfest's theorem, which states that

$$\begin{cases} \langle F\rangle = \frac{d\langle p\rangle}{dt} \\ \langle p\rangle = m\frac{d\langle x\rangle}{dt} \end{cases}, \tag{1.109}$$

establishing a perfect analogy between classical and quantum mechanics, in the sense that quantum expectation fulfills classical Newtonian mechanics.

To prove the previous relations, we consider both

$$\frac{d\langle p\rangle}{dt} = \frac{i}{\hbar}\left\langle\left[\frac{p^2}{2m} + V(x), p\right]\right\rangle \tag{1.110}$$

and

$$\frac{d\langle x\rangle}{dt} = \frac{i}{\hbar}\left\langle\left[\frac{p^2}{2m} + V(x), x\right]\right\rangle, \tag{1.111}$$

respectively, where the fact that position and momentum operators do not have time dependences was taken into account. By developing the first relation, taking into account that the momentum commutes with itself, and substituting the momentum with $p \to -i\hbar\frac{\partial}{\partial x}$, it results that

$$\frac{d\langle p\rangle}{dt} = \int_{\mathbb{R}} \overline{\Psi} V(x)\frac{\partial \Psi}{\partial x}\, dx - \int_{\mathbb{R}} \overline{\Psi}\frac{\partial}{\partial x}\left(V(x)\Psi\right)\, dx. \tag{1.112}$$

Developing the second integral in the previous equation, it ultimately, results that

$$\frac{d\langle p\rangle}{dt} = -\int_{\mathbb{R}} \overline{\Psi}\frac{\partial V(x)}{\partial x}\Psi\, dx, \tag{1.113}$$

or

$$\frac{d\langle p\rangle}{dt} = -\left\langle \frac{\partial V(x)}{\partial x}\right\rangle = \langle F\rangle, \tag{1.114}$$

which generalizes Newton's second law.

Now, coming back to the second expression in Eq. (1.109), and by using the commutation relation (which again reflects the Heisenberg uncertainty principle)

$$[p, x] = -i\hbar, \tag{1.115}$$

it results that

$$\begin{aligned} \frac{d\langle x\rangle}{dt} &= \frac{i}{\hbar}\left\langle\left[\frac{p^2}{2m} + V(x), x\right]\right\rangle = \frac{i}{\hbar}\left\langle\left[\frac{p^2}{2m}, x\right]\right\rangle \\ &= \frac{i}{2m\hbar}\left\langle p\,[p, x] + [p, x]p\right\rangle = \frac{1}{m}\langle p\rangle. \end{aligned} \tag{1.116}$$

Again, all the previous results can be easily extended to the 3D case.

1.9 Charge Density and Interatomic Potentials: The Hellmann–Feynman Theorem

As soon as the electronic distribution is known, the electron density $\rho_e(\mathbf{r})$ can be easily computed by applying

$$\rho_e(\mathbf{r}) = \int\limits_{\mathbb{R}^{3(N_e-1)}} \overline{\Psi}\cdot\Psi\, d\mathbf{r}_1 \ldots d\mathbf{r}_{e-1} d\mathbf{r}_{e+1} \ldots d\mathbf{r}_{N_e}, \tag{1.117}$$

allowing us to compute the electronic density at each point within the space by adding all of the electron contributions according to

$$\rho(\mathbf{r}) = \sum_{e=1}^{N_e} \rho_e(\mathbf{r}). \tag{1.118}$$

Now, the Hellmann–Feynman theorem [9] leads to a direct interpretation of the electronic effects on the nuclei in terms of classical electrostatics. The applied force on a nucleus k results in

$$\mathbf{f}_k = -\nabla_k(\widetilde{V}_{nk} + \widetilde{V}_{ek}), \tag{1.119}$$

where $\widetilde{V}_{nk}$ is the electrostatic inter-nuclei contribution

$$\widetilde{V}_{nk} = \sum_{n=1, n\neq k}^{N_n} \frac{Z^2}{\|\mathbf{R}_n - \mathbf{R}_k\|}, \tag{1.120}$$

and $\widetilde{V}_{ek}$ is the one associated with the electronic distribution:

$$\widetilde{V}_{ek} = -Z\left(\int_{\mathbb{R}^3} \frac{\rho(\mathbf{r})}{\|\mathbf{r} - \mathbf{R}_k\|} d\mathbf{r}\right). \tag{1.121}$$

Thus, it seems clear that by knowing the electronic distribution, the electronic density can be computed, and from it, the effects that the electronic distribution has on the nuclei. The solution of the Schrödinger equation allows for computing the exact interatomic potentials that could then be used to perform accurate molecular dynamics simulations. However, the multi-dimensional character of the Schrödinger equation makes such a solution very difficult, even for moderate populations of electrons. Two possibilities exist: (i) the one that consists in introducing some simplifications into the treatment of that equation; or (ii) the one related to the proposal of empirical (or quantum-inspired) inter-atomic potentials to be used in the molecular dynamics framework, as discussed in Chap. 3. Ab initio simulations are performed using quantum-based potential, whereas molecular dynamics uses empirical potentials.

Before analyzing the state-of-the-art of the computational treatment of the Schrödinger equation, we are introducing a final key concept, the one related to the Pauli exclusion principle.

1.10 The Pauli Exclusion Principle

The wavefunction of many-electrons must reflect the fact that electrons are indistinguishable. Electrons do not know that we have labeled them. If we use $\mathbf{r}_1$ to describe the coordinates of one electron, and $\mathbf{r}_2$ for another, then

$$|\Psi(\mathbf{r}_1, \mathbf{r}_2, \mathbf{r}_3, \ldots, \mathbf{r}_{Ne})|^2 = |\Psi(\mathbf{r}_2, \mathbf{r}_1, \mathbf{r}_3, \ldots, \mathbf{r}_{N_e})|^2, \qquad (1.122)$$

that is, the probability distribution of electrons cannot depend on the way that we have labeled them. Thus, if Π is any permutation of the N_e electronic coordinates, then

$$\Pi|\Psi(\mathbf{r}_1, \mathbf{r}_2, \mathbf{r}_3, \ldots, \mathbf{r}_{Ne})|^2 = |\Psi(\mathbf{r}_1, \mathbf{r}_2, \mathbf{r}_3, \ldots, \mathbf{r}_{N_e})|^2, \qquad (1.123)$$

which implies just two possibilities,

$$\Pi\Psi(\mathbf{r}_1, \mathbf{r}_2, \mathbf{r}_3, \ldots, \mathbf{r}_{Ne}) = \Psi(\mathbf{r}_1, \mathbf{r}_2, \mathbf{r}_3, \ldots, \mathbf{r}_{N_e}), \qquad (1.124)$$

or

$$\Pi\Psi(\mathbf{r}_1, \mathbf{r}_2, \mathbf{r}_3, \ldots, \mathbf{r}_{Ne}) = -\Psi(\mathbf{r}_1, \mathbf{r}_2, \mathbf{r}_3, \ldots, \mathbf{r}_{N_e}), \qquad (1.125)$$

which can be written as

$$\Pi\Psi(\mathbf{r}_1, \mathbf{r}_2, \mathbf{r}_3, \ldots, \mathbf{r}_{Ne}) = a\Psi(\mathbf{r}_1, \mathbf{r}_2, \mathbf{r}_3, \ldots, \mathbf{r}_{N_e}), \qquad (1.126)$$

where the exclusion Pauli principle states that, for electrons, $a = -1$.

Thus, the most general statement of the Pauli principle for electrons establishes that an acceptable wavefunction for many electrons must be antisymmetric with respect to the exchange of the coordinates of any two electrons.

If we imagine a system composed of two non-interacting electrons ($V_{e_1e_2} = 0$), the resulting wavefunction can be written in the one-term separated form

$$\Psi(\mathbf{r}_1, \mathbf{r}_2) = \Psi_1(\mathbf{r}_1) \cdot \Psi_2(\mathbf{r}_2), \qquad (1.127)$$

but because of the Pauli exclusion constraint, this function must be adjusted to ensure its antisymmetry ($a = -1$ in Eq. (1.126)) by considering the Slater determinant

$$\Psi(\mathbf{r}_1, \mathbf{r}_2) = \Psi_1(\mathbf{r}_1) \cdot \Psi_2(\mathbf{r}_2) - \Psi_1(\mathbf{r}_2) \cdot \Psi_2(\mathbf{r}_1) = \begin{vmatrix} \Psi_1(\mathbf{r}_1) & \Psi_2(\mathbf{r}_1) \\ \Psi_1(\mathbf{r}_2) & \Psi_2(\mathbf{r}_2) \end{vmatrix}, \qquad (1.128)$$

where use of the determinant ensures that changing the electron's label corresponds to exchanging the two associated columns in the determinant, and the associated change in the wavefunction sign. However, this strategy introduces some technical difficulties by increasing the number of terms involved in the wavefunction expression.

Until now, we have not included anything in the description that takes into account the electron spin. One could think that, as there is no mention of the spin in the Hamiltonian, the spin should have little relevance to the energy of the system. However, the Pauli principle applies to the exchange of all of the coordinates (space and spin), so in order to comply with it, the spin must be introduced into the particle coordinates.

If we do not introduce the spin into the description of a system composed of two electrons, the solution must verify that $\Psi(\mathbf{r}_1, \mathbf{r}_2) = -\Psi(\mathbf{r}_2, \mathbf{r}_1)$, but if the spin is considered, the expression $\Psi(\mathbf{r}_1, \mathbf{r}_2) = \Psi(\mathbf{r}_2, \mathbf{r}_1)$ holds if the two electrons have opposite spin.

We can give a simple solution to this issue by including the spin in the particles' coordinates by defining the extended coordinates $\mathbf{x}_i$ from

$$\mathbf{x}_i = (\mathbf{r}_i, s_i) = (x_i, y_i, z_i, s_i), \tag{1.129}$$

where s_i is the spin coordinate of electron i, which can take two values that we denote by α and β (in fact, $-1/2$ and $1/2$ in the case of electrons). Thus, two electrons can be allocated in the same orbital as soon as they have different spin. The Pauli principle is of major importance for explaining the electronic structure of matter, and also atomic bonding.

Remark. When one is solving the transient Schrödinger equation, if the initial condition verifies the Pauli exclusion principle, the computed transient solution will verify the antisymmetry without considering any particular treatment. However, when the time-independent solution is searched, the antisymmetry must be enforced, and the most common way to enforce it is the use of Slater's determinant. It is important to note that the computation of transient solutions needs very efficient integration schemes able to ensure the energy conservation while guaranteeing the stability.

1.11 On the Numerical Solution of the Schrödinger Equation

Due to the high dimensionality of the Schrödinger equation, its solution is only possible for very reduced populations of electrons. For this reason, different approximated methods have been proposed and extensively used. We describe in the next chapter the two most widely used: the Hartree–Fock (HF) method and the Density Functional Theory (DFT), the first of which can be applied only for a moderate number of electrons and it has been extensively used in the context of quantum chemistry to analyze the structure and behavior of molecules. However, in the case of crystals, the number of electrons becomes too large to make their simulation possible by the direct solution of the Schrödinger equation or by using the Hartree–Fock technique. In this context, the most successful technique was and continues to be the DFT. Both techniques (HF and DFT) are, in practice, approximated techniques that work in some cases and could become crude approximations in other cases.

The three techniques (direct Schrödinger solution, HF and DFT) can be applied for solving both the time-independent and the transient Schrödinger equations. Transient solutions are very common in the context of quantum gas dynamics (physics of plasma), but they are more scarce in material science when the structure and properties of molecules or crystals are concerned. For this reason, in that which follows we are focusing on the solution of the time-independent Schrödinger equation that leads to the solution of its associated multidimensional eigenproblem.

Before considering the two approximated techniques (HF and DFT), in the next chapter, we consider here the most natural and accurate one, the one consisting in the direct solution of the Schrödinger equation. In the time-independent case, usually only the so-called ground distribution (the one related to the minimum energy) is searched. In any case, the high dimensional spaces in which the Schrödinger equation is defined leads to the curse of dimensionality. Nowadays, it is widely accepted that classical discretization methods, based on the use of a grid (or a mesh), fail when the space dimension approaches a few dozen. Some attempts to address multidimensional equations have been proposed: the simplest choice consists in using stochastic techniques (described in Sect. 3.2), because in this case, the computation complexity does not scale exponentially with the space dimension. However, these simulations are expensive and introduce statistical noise. Moreover, only the moments of the distribution can be computed accurately, because an accurate description of the distribution itself requires too many trajectories of the stochastic process. A second possibility consists in using sparse grid techniques, but nowadays, the use of these techniques only allows for solving models in spaces rarely exceeding a few dozen. A third possibility consists in employing separated representations (at the origin of the so-called Proper Generalized Decomposition) that allow for an accurate and fast solution of models involving many particles [11]. Even if the use of separated representations in computational mechanics is quite unusual, its use has been extensively considered in the context of quantum chemistry; in particular, it is at the base of the Hartree–Fock method described in the next chapter. Within the context of separated representations, the main issue when addressing the solution of the Schrödinger equation is related to the antisymmetry of the wavefunction enforced by using the Slater determinant, whose number of terms explodes with the space dimension (number of particles involved in the quantum system).

In order to illustrate the limits of classical discretization techniques for addressing the direct solution of the Schrödinger equation, we consider two quotations extracted from [12]:

For the solution of the time-dependent problems, different levels of approximation have been used, which range between the direct discretization of the TDSE – time dependent Schrödinger equation – as the most precise but computationally most expensive choice, and time-dependent density functional theory (TDDFT), which is appealing from a computational point of view but turned out to be too crude an approximation to capture important features of the problem, ...

The present state-of-the-art of numerically solving the time-dependent Schrödinger equation directly for realistic laser pulses is limited to two electron systems. The most successful calculations involve the largest massively parallel computers available. It is clear that the direct solution of the linear time-dependent Schrödinger equation has reached its computational limits;

or the analysis addressed in [13]:

... the theory of Everything is not even remotely a theory of every thing. We know this equation [the Schrödinger equation] is correct because it has been solved accurately for a small number of particles -isolated atoms and small molecules - and found to agree in minute detail with experiments. However, it cannot be solved accurately when the number of particles exceeds about 10. No computer in existence, or that will ever exist, can break this barrier because it is a catastrophe of dimension. If the amount of computer memory required to represent the quantum wavefunction of one particle is N, then the amount required to represent the wavefunction of k particles is N^k. *It is possible to perform approximate calculations ... but the schemes for approximating are not first-principles deductions but rather are keyed to experiments ...*

One could surmise that this pessimistic perspective is based on the fact that the complexity scaling with N^k is due to the fact that a grid was used instead of separated representations. In some cases, the use of a separated representation circumvents the curse of dimensionality, leading to a complexity scaling with $N \cdot k$ instead of N^k, however, some quantum systems explore the whole configurational space whose dimension corresponds to N^k (for example, the spin glass problem considered in [14] consisting of a set of k quantum spins interacting by random Heisenberg exchanges where the configuration space dimension becomes 2^k and the system can obviously explore the whole configuration space). In these cases, no solution exists! Thus, we must assume the existence of numerically tractable and intractable quantum systems. In what follows, we are considering only the first ones and addressing their solutions by using approximated techniques.

References

1. E. Wichmann, *Quantum Physics* (McGraw-Hill, New York, 1971)
2. S. Ortoli, J.P. Pharabod, *Le cantique des quantiques* (La Dcouverte, 1984)
3. J. Gribbin, *In Search of Schrdingers Cat* (Bantam Books, New York, 1984)
4. B. Diu, *Trait de physique lusage des profanes* (Odile Jacob, 2000)
5. A. Rae, *Quantum Physics: Illusion or Reality?* (Cambridge University Press, Cambridge, 2004)
6. J.L. Basdevant, *12 Leçons de mcanique quantique* (Vuibert, 2006)
7. R. Omnès, *Les indispensables de la mcanique quantique* (Odile Jacob, 2006)
8. J. Hladik, *Pour comprendre simplement les origines de lvolution de la Physique quantique* (Ellipses, 2007)
9. J.H. Weiner, *Statistical Mechanics of Elasticity* (Dover, New York, 2002)
10. D.B. Cook, *Handbook of Computational Chemistry* (Oxford University Press, New York, 1998)
11. A. Ammar, F. Chinesta, P. Joyot, The nanometric and micrometric scales of the structure and mechanics of materials revisited: an introduction to the challenges of fully deterministic numerical descriptions. Int. J. Multiscale Comput. Eng. **6**(3), 191–213 (2008)
12. O. Koch, W. Kreuzer, A. Scrinzi, Approximation of the time-dependent electronic Schrödinger equation by MCTDHF. Appl. Math. Comput. **173**, 960–976 (2006)
13. R.B. Laughlin, The theory of everything. Proc. USA Natl. Acad. Sci. (2000)
14. B.A. Bernevig, D. Giuliano, R.B. Laughlin, Coordinate representation of the one-spinon one-holon wavefunction and spinon-holon interaction. Phys. Rev. B **65**, 195112 (2002)

Chapter 2
Ab-Initio Calculations

Abstract Due to the difficulties found in the direct solution of the Schrödinger equation, different simplified approaches were proposed and are nowadays widely used. Among them, those most usually employed are the Hartree–Fock and the Density Functional Theory, which we revisit in the present chapter. The former makes use of nonstandard numerical approximations in order to calculate the wavefunction while circumventing the curse of dimensionality, whereas the latter involves the electronic density that is now defined in three dimensions but requires deeper analyses to retain the most relevant features present in the wavefunction description in a coarse 3D model.

Keywords Hartree-Fock · Hohenberg and Kohn Theorems · Density Functional Theory

2.1 The Hartree–Fock Description

2.1.1 The Orbital Model

The set of all the solutions to the one-electron Schrödinger equation reads as

$$\hat{\mathscr{H}}\, \phi_i = \hat{E}_i\, \phi_i, \tag{2.1}$$

where $\hat{\mathscr{H}}$ represents the one-electron Hamiltonian hermitian operator. The eigenfunctions ϕ_i, known as spatial orbitals, related to the eigenvalues $\hat{E}_i$ (energies), define a complete basis of the 3D space, such that any 3D function can be written as

$$f(\mathbf{r}) = \sum_{j=1}^{\infty} c_j\, \phi_j(\mathbf{r}), \tag{2.2}$$

© The Author(s) 2018

F. Chinesta and E. Abisset-Chavanne, *A Journey Around the Different Scales Involved in the Description of Matter and Complex Systems*, SpringerBriefs in Applied Sciences and Technology, https://doi.org/10.1007/978-3-319-70001-4_2

where **r** denotes the space coordinates, after eliminating the spin coordinate in expression (1.129), i.e., $\mathbf{r} = (x, y, z)$.

If we define the spin-orbitals $\varphi_j(\mathbf{x})$ as

$$\varphi_j(\mathbf{x}) = \phi_j(\mathbf{r}) \cdot \alpha(s), \tag{2.3}$$

or

$$\varphi_j(\mathbf{x}) = \phi_j(\mathbf{r}) \cdot \beta(s), \tag{2.4}$$

then the solution $\Psi(\mathbf{x}_1, \mathbf{x}_2)$ of the two-electrons Schrödinger equation could be approximated as follows: fixing the value of one of the coordinates, e.g., $\mathbf{x}_2$, and using the rationale just described, it results that

$$\Psi(\mathbf{x}_1; \mathbf{x}_2) = \sum_{j=1}^{\infty} c_j(\mathbf{x}_2)\, \varphi_j(\mathbf{x}_1), \tag{2.5}$$

and considering

$$c_j(\mathbf{x}_2) = \sum_{k=1}^{\infty} d_k^j\, \varphi_k(\mathbf{x}_2), \tag{2.6}$$

it ultimately results that

$$\Psi(\mathbf{x}_1, \mathbf{x}_2) = \sum_{j=1}^{\infty} \sum_{k=1}^{\infty} c_{jk}\, \varphi_j(\mathbf{x}_1) \cdot \varphi_k(\mathbf{x}_2), \tag{2.7}$$

where $c_{jk} = d_k^j$.

This expression can be generalized to the many-electrons distribution function.

2.1.2 *Accounting for the Pauli Exclusion Principle*

In order to ensure verification of the Pauli exclusion principle, we define the determinants

$$\Phi_k(\mathbf{x}_1, \cdots, \mathbf{x}_{N_e}) = \begin{vmatrix} \varphi_{m_1^k}(\mathbf{x}_1) & \cdots & \varphi_{m_1^k}(\mathbf{x}_{N_e}) \\ \vdots & \ddots & \vdots \\ \varphi_{m_{N_e}^k}(\mathbf{x}_1) & \cdots & \varphi_{m_{N_e}^k}(\mathbf{x}_{N_e}) \end{vmatrix}, \tag{2.8}$$

where k refers to a particular choice of the N_e indexes $m_1^k, \cdots, m_{N_e}^k$.

Thus, the multi-electronic wavefunction can be approximated as

$$\Psi(\mathbf{x}_1, \cdots, \mathbf{x}_{N_e}) = \sum_{k=1}^{\infty} D_k\, \Phi_k(\mathbf{x}_1, \cdots, \mathbf{x}_{N_e}). \tag{2.9}$$

A permutation in the label of two electrons implies the exchange of two columns of the different determinants involved in Eq. (2.9), and then a change of the sign in agreement with Pauli exclusion principle.

2.1.3 The Variational Principle

In order to compute the wavefunction approximate given by Eq. (2.9), we must prove the existence of a variational principle whose minimization will result in the desired wavefunction.

The eigenproblem related to the multi-electronic system reads as

$$\mathscr{H}\,\Psi = E\,\Psi, \tag{2.10}$$

which results in the eigenfunctions Ψ_i verifying the orthonormality condition

$$\int \overline{\Psi_i}\,\Psi_j\,d\mathbf{r} = \delta_{ij}, \tag{2.11}$$

where $d\mathbf{r} = d\mathbf{r}_1 \cdot d\mathbf{r}_2 \cdots d\mathbf{r}_{N_e}$. It is important to note that since the Hamiltonian is independent of the spin, the resulting eigenfunctions only depend on the space coordinates.

Although Ψ_i are and remain unknown, their formal properties ensure that they form a complete basis for the expression of any function. Thus, if we write

$$\Psi' = \sum_{j=1}^{\infty} B_j\,\Psi_j, \tag{2.12}$$

where B_j are arbitrary coefficients, then the associated energy (according to (Eq. 1.83)), results in

$$E' = \frac{\int \overline{\Psi'}\mathscr{H}\Psi' d\mathbf{r}}{\int \overline{\Psi'}\Psi' d\mathbf{r}}, \tag{2.13}$$

where the denominator accounts for the non-normality of Ψ'.

Introducing the approximation (2.12) into the expression (2.13) and taking into account Eq. (2.11), it results that

$$E' = \frac{\sum_{j=1}^{\infty} |B_j|^2 E_j}{\sum_{j=1}^{\infty} |B_j|^2}, \tag{2.14}$$

which subtracting the lowest energy E_1 (ground state), yields

$$(E' - E_1) = \frac{1}{\sum_{j=1}^{\infty} |B_j|^2} \sum_{j=1}^{\infty} |B_j|^2 (E_j - E_1) \geq 0, \tag{2.15}$$

implying that

$$E' \geq E_1, \tag{2.16}$$

which means that whatever function of N_e electronic coordinates one chooses, the mean value of the Hamiltonian operator is always greater than the lowest true energy of the associated Schrödinger equation, giving the procedure for finding numerical solutions. It suffices to minimize the Hamiltonian operator applied to the test wavefunction approximate as described in the next section.

2.1.4 A Direct Solution Procedure

If the expression of Ψ' is written as a linear combination of a finite number M of determinants, i.e.,

$$\Psi'(\mathbf{x}_1, \ldots, \mathbf{x}_{N_e}) = \sum_{j=1}^{M} D_j \, \Phi_j(\mathbf{x}_1, \ldots, \mathbf{x}_{N_e}), \tag{2.17}$$

then the associated energy results in

$$E' = \frac{\sum_{j=1}^{M} \sum_{k=1}^{M} \overline{D_j} D_k \int \overline{\Phi_j} \mathscr{H} \Phi_k d\mathbf{r}}{\sum_{j=1}^{M} \sum_{k=1}^{M} \overline{D_j} D_k \int \overline{\Phi_j} \Phi_k d\mathbf{r}}. \tag{2.18}$$

Introducing the notation

$$\begin{cases} H_{jk} = \int \overline{\Phi_j} \mathscr{H} \Phi_k \, d\mathbf{r} \\ S_{jk} = \int \overline{\Phi_j} \Phi_k \, d\mathbf{r} \end{cases}, \tag{2.19}$$

Equation (2.18) reads as

$$\left(\sum_{j=1}^{M} \sum_{k=1}^{M} \overline{D_j} D_k S_{jk} \right) E' = \sum_{j=1}^{M} \sum_{k=1}^{M} \overline{D_j} D_k H_{jk}, \tag{2.20}$$

whose minimization leads to

$$\left(\sum_{k=1}^{M} D_k S_{jk}\right) E = \sum_{k=1}^{M} D_k H_{jk}, \quad \forall j, \tag{2.21}$$

which can be written in the matrix form

$$\begin{pmatrix} H_{11} & \dots & H_{1M} \\ \vdots & \ddots & \vdots \\ H_{M1} & \dots & H_{MM} \end{pmatrix} \begin{pmatrix} D_1 \\ \vdots \\ D_M \end{pmatrix} = E \begin{pmatrix} S_{11} & \dots & S_{1M} \\ \vdots & \ddots & \vdots \\ S_M & \dots & S_{MM} \end{pmatrix} \begin{pmatrix} D_1 \\ \vdots \\ D_M \end{pmatrix}, \tag{2.22}$$

or

$$\mathbf{H}\,\mathbf{D} = E\,\mathbf{S}\,\mathbf{D}. \tag{2.23}$$

The main difficulties in this numerical approach are as follows:

- How many determinants M should be considered in the expansion (2.12)?
- How to quantify the solution quality?
- What are the most appropriate spin-orbitals $\varphi_i(\mathbf{x})$ for performing the development?
- What are the best determinants to consider, that is, the best choices of indices m_i^k, $\forall k$?
- Despite the fact that coefficients H_{jk} and S_{jk} are known, in principle, because everything is known about the integrals, they still remain formidable technical problems, being integrals of $3N_e$ spatial variables. The integrals can be separated as a sum of products of integrals defined in the 3D spaces. However, the separated form of integrals H_{jk} requires the integration in 6D spaces, because of the electron-electron potential that appears in the Hamiltonian.

2.1.5 The Hartree–Fock Approach

As the general expansion (2.17) is computationally too expensive, one could try to capture the main features of the solution by assuming that this sum reduces to a single term. Of course, if one uses

$$\Psi'(\mathbf{x}_1, \dots, \mathbf{x}_{N_e}) = D\,\Phi(\mathbf{x}_1, \dots, \mathbf{x}_{N_e}), \tag{2.24}$$

there would be no chance of defining an acceptable solution, except by considering that the determinant $\Phi(\mathbf{x}_1, \dots, \mathbf{x}_{N_e})$ is defined from a set of separate unknown orbitals that should be computed from the minimization that the variational principle imposes. Thus, the Hartree–Fock approach considers the Hartree–Fock wavefunction $\Phi^{HF}(\mathbf{x}_1, \dots, \mathbf{x}_{N_e})$ defined from

$$\Phi^{HF}(\mathbf{x}_1, \ldots, \mathbf{x}_{N_e}) = \begin{vmatrix} \chi_{m_1}(\mathbf{x}_1) & \ldots & \chi_{m_1}(\mathbf{x}_{N_e}) \\ \vdots & \ddots & \vdots \\ \chi_{m_{N_e}}(\mathbf{x}_1) & \ldots & \chi_{m_{N_e}}(\mathbf{x}_{Ne}) \end{vmatrix}, \tag{2.25}$$

in which, as just indicated, orbitals χ_i are approximated from m spin-orbitals (previously introduced) according to

$$\chi_i(\mathbf{x}) = \sum_{r=1}^{m} C_{ri}\ \varphi_r(\mathbf{x}). \tag{2.26}$$

Coefficients C_{ri} are computed by using the variational formulation associated with the energy

$$E\left[\Phi^{HF}\right] = \frac{\int \overline{\Phi}^{HF} \mathscr{H} \Phi^{HF} d\mathbf{r}}{\int \overline{\Phi}^{HF} \Phi^{HF} d\mathbf{r}}, \tag{2.27}$$

where $E\left[\Phi^{HF}\right]$ indicates that the energy is a functional of the Hartree–Fock wavefunction, which could also be written as $E[\chi_i]$. The interested reader can refer to [1] for additional details on the calculation procedure.

2.1.6 Post-Hartree–Fock Methods

Note that Eq. (2.26) involves $N_e \cdot m$ unknown complex coefficients. Thus, the computational complexity scales in $N_e \cdot m$, that is, linearly with the dimension of the space (number of electrons N_e) or with the number m of functions used in the approximation of the orbitals χ_i. This scalability is characteristic of separated representations [2].

The main limitation of the Hartree–Fock method lies in the single-determinant expansion used in the approximation of the wavefunction solution of the multi-electronic Schrödinger equation. If the main features present in this solution cannot be expressed from a single-determinant expansion, the Hartree–Fock solution could be inaccurate.

To circumvent this crude approximation, different multi-determinant approaches have been proposed. Interested readers can refer to [3], as well as to the different chapters of the handbook on computational chemistry [4]. The simplest alternative consists in writing the solution as a linear combination of some Slater determinants, built by combining m orbitals, with $m > N_e$. These orbitals are assumed as being known (e.g., the orbitals related to the hydrogen atom) and the weights are searched to minimize the electronic energy. When the molecular orbitals are built from the Hartree–Fock solution (by employing the ground state and some excited eigenfunctions), the technique is known as the Configuration Interaction method (CI).

A more sophisticated technique consists in writing this multi-determinant approximation of the solution by using a number of molecular orbitals m (with $m > N_e$) that are assumed to be unknown. Thus, the minimization of the electronic energy leads to simultaneous computation of the molecular orbitals and the associated coefficients of this multi-determinant expansion. Obviously, each one of these unknown molecular orbitals is expressed in an appropriate functional basis (e.g., Gaussian functions, ...). This strategy is known as a Multi-Configuration Self-Consistent Field (MCSCF).

All of the just-mentioned strategies (and others like the coupled cluster or the Moller–Plesset perturbation methods) belong to the family of wavefunction-based methods. They can only be used to solve quantum systems composed of a moderate number of electrons, because the number of terms involved in the determinants scales with the factorial of the number of electrons, i.e., with $N_e!$ (the factorial of N_e).

2.2 Density Functional Theory

Solid physics deals with multi-electron systems implying billions of particles, not just dozens as in molecular theories. This means that methods based on electron density are much more widely used. The constant efforts to develop such methods have been rewarded by a series of amazing theorems showing that it is possible to obtain the exact electron density without using the wavefunction.

Density functional theory–DFT–is based on two major results, the so-called Hohenberg and Kohn theorems.

2.2.1 The First Hohenberg and Kohn Theorem

The first Hohenberg and Kohn theorem states that the electronic density uniquely determines the external potential, the one created by the nuclei.

We start by defining the electronic density in the context of the single-determinant approach (which implies operating with space-spin coordinates)

$$\rho(\mathbf{r}) = N_e \int \overline{\Phi}(\mathbf{x}, \mathbf{x}_2, \ldots, \mathbf{x}_{N_e}) \Phi(\mathbf{x}, \mathbf{x}_2, \ldots, \mathbf{x}_{N_e})\, d\mathbf{r}_2 \ldots d\mathbf{r}_{N_e}, \tag{2.28}$$

or

$$\rho(\mathbf{r}) = N_e \int |\Psi(\mathbf{x}, \mathbf{x}_2, \ldots, \mathbf{x}_{N_e})|^2\, d\mathbf{r}_2 \ldots d\mathbf{r}_{N_e}. \tag{2.29}$$

We assume the following Hamiltonian partition:

$$\mathcal{H} = \mathcal{T} + \mathcal{V} + \mathcal{G}, \tag{2.30}$$

where $\mathscr{T}$ represents the kinetic energy operator, $\mathscr{V}$ the external potential operator (the one created by the nuclei) and $\mathscr{G}$ the inter-electron repulsions potential.

In order to prove that the electronic density uniquely determines the external potential, we assume that two different external potentials correspond to the same electronic density. This fact implies different Hamiltonians that only differ due to the difference in the external potentials because the kinetic energy part and the one corresponding to the inter-electron interactions are the same as soon as the number of electrons is the same. We denote the two different external potentials by $\mathscr{V}$ and $\mathscr{V}'$, and their corresponding Hamiltonians by $\mathscr{H}$ and $\mathscr{H}'$, respectively. As the Hamiltonian determines the wavefunction, these will be denoted by Ψ and Ψ'.

Now, the variational principle introduced in Sect. 2.1.3 states

$$\begin{cases} \int \overline{\Psi'} \mathscr{H} \Psi' \, d\mathbf{r}_1 \cdots d\mathbf{r}_{N_e} > E \\ \int \overline{\Psi} \mathscr{H}' \Psi \, d\mathbf{r}_1 \cdots d\mathbf{r}_{N_e} > E' \end{cases}, \tag{2.31}$$

with

$$\begin{cases} E = \int \overline{\Psi} \mathscr{H} \Psi \, d\mathbf{r}_1 \cdots d\mathbf{r}_{N_e} \\ E' = \int \overline{\Psi'} \mathscr{H}' \Psi' \, d\mathbf{r}_1 \cdots d\mathbf{r}_{N_e} \end{cases}. \tag{2.32}$$

Thus, considering the first expression in Eq. (2.31), it results that

$$E < \int \overline{\Psi'} \mathscr{H} \Psi' \, d\mathbf{r}_1 \cdots d\mathbf{r}_{N_e} =$$

$$\int \overline{\Psi'} \mathscr{H}' \Psi' \, d\mathbf{r}_1 \cdots d\mathbf{r}_{N_e} + \int \overline{\Psi'} (\mathscr{H} - \mathscr{H}') \Psi' \, d\mathbf{r}_1 \cdots d\mathbf{r}_{N_e} =$$

$$E' + \int \overline{\Psi'} (\mathscr{V} - \mathscr{V}') \Psi' \, d\mathbf{r}_1 \cdots d\mathbf{r}_{N_e} = E' + \int (v(\mathbf{r}) - v'(\mathbf{r})) \rho(\mathbf{r}) \, d\mathbf{r}, \tag{2.33}$$

where $v(\mathbf{r})$ refers to the one-electron potential (see Sect. 1.9).

Now, applying the same rationale to the second expression in Eq. (2.31), we obtain

$$E' < \int \overline{\Psi} \mathscr{H}' \Psi \, d\mathbf{r}_1 \cdots d\mathbf{r}_{N_e} =$$

$$\int \overline{\Psi} \mathscr{H} \Psi \, d\mathbf{r}_1 \cdots d\mathbf{r}_{N_e} + \int \overline{\Psi} (\mathscr{H}' - \mathscr{H}) \Psi \, d\mathbf{r}_1 \cdots d\mathbf{r}_{N_e} =$$

$$E + \int \overline{\Psi} (\mathscr{V}' - \mathscr{V}) \Psi \, d\mathbf{r}_1 \cdots d\mathbf{r}_{N_e} =$$

$$E + \int (v'(\mathbf{r}) - v(\mathbf{r})) \rho(\mathbf{r}) \, d\mathbf{r} = E - \int (v(\mathbf{r}) - v'(\mathbf{r})) \rho(\mathbf{r}) \, d\mathbf{r}. \tag{2.34}$$

By adding Eqs. (2.33) and (2.34), it results that

$$E' + E < E' + E, \tag{2.35}$$

from which we conclude that $\mathscr{V}' = \mathscr{V}$ and $\mathscr{H}' = \mathscr{H}$. Since the wavefunction depends on the Hamiltonian, we can affirm that the wavefunction is uniquely determined by the electron density, and consequently $\Psi' = \Psi$.

2.2.2 The Second Hohenberg and Kohn Theorem

Now, in order to determine the electronic density, the second Hohenberg and Kohn theorem establishes a variational principle whose minimization results in the desired electronic distribution.

From a given electronic density $\rho'(\mathbf{r})$, we can write

$$E' = \int \overline{\Psi'} \mathscr{H} \Psi' d\mathbf{r}_1 \cdots d\mathbf{r}_{N_e} = W[\rho'(\mathbf{r})]. \tag{2.36}$$

The variational principle introduced in Sect. 2.1.3 implies that

$$E' = W[\rho'(\mathbf{r})] \geq E = \int \overline{\Psi} \mathscr{H} \Psi d\mathbf{r}_1 \cdots d\mathbf{r}_{N_e} = W[\rho(\mathbf{r})], \tag{2.37}$$

with $\mathscr{H}\Psi = E\Psi$, which establishes the desired result

$$\begin{cases} W[\rho'(\mathbf{r})] \geq W[\rho(\mathbf{r})] \\ \int \rho'(\mathbf{r}) d\mathbf{r} = N_e \end{cases} . \tag{2.38}$$

2.2.3 The Hohn–Sham Equations

In the density functional theory, two conceptual difficulties remain:

- How to quantify the electron's kinetic energy solely with the knowledge of their distributions in space?
- What is the role of antisymmetry (Pauli exclusion principle) requirements in the electron density function?

We start by approximating the unknown function, the trial density, within a single-determinant approach

$$\rho(\mathbf{r}) = N_e \int \overline{\Phi}(\mathbf{x}, \mathbf{x}_2, \ldots, \mathbf{x}_{N_e}) \Phi(\mathbf{x}, \mathbf{x}_2, \ldots, \mathbf{x}_{N_e}) \, d\mathbf{r}_2 \ldots d\mathbf{r}_{N_e}, \tag{2.39}$$

where the determinant Φ involves the unknown orthonormal spin-orbitals $\chi_i(\mathbf{x}_i)$, an approximation that is in agreement with the Pauli exclusion principle and that verifies the N_e-representability

$$\int \rho(\mathbf{r}) d\mathbf{r} = N_e. \tag{2.40}$$

The total energy $W[\rho(\mathbf{r})]$ can be decomposed into three contributions, one related to the kinetic energy $T[\rho(\mathbf{r})]$, one that considers the external potential (electron-nuclei interactions) $V[\rho(\mathbf{r})]$ and finally one representing the electron-electron interactions $J[\rho(\mathbf{r})]$, with the last two contributions reading as

$$V[\rho(\mathbf{r})] = \int v(\mathbf{r})\rho(\mathbf{r}) d\mathbf{r}, \tag{2.41}$$

and

$$J[\rho(\mathbf{r})] = \int \rho(\mathbf{r}_1) \frac{1}{||\mathbf{r}_1 - \mathbf{r}_2||} \rho(\mathbf{r}_2) d\mathbf{r}_1 d\mathbf{r}_2. \tag{2.42}$$

For the kinetic energy, we assume an initial contribution $\hat{T}$ by assuming non-interacting electrons

$$\hat{T}[\rho(\mathbf{r})] = \sum_{i=1}^{N_e} \int \overline{\chi_i} \left(\frac{1}{2} \nabla^2 \right) \chi_i \, d\mathbf{r}_i, \tag{2.43}$$

where χ_i are the orbitals considered in the expression of the determinant Φ.

The remaining contribution to the kinetic energy and the non-Coulomb effects are grouped in the exchange-correlation-residual-kinetic energy $E_{XCKE}[\rho(\mathbf{r})]$. The main difficulty concerns the expression of the exchange-correlation-kinetic-residual energy that is not known. In general, this term is obtained through a combination of heuristic arguments, because more accurate techniques exploiting the self-consistency are too expensive to be used.

Now, the solution procedure consists of the following steps

- Associate a variation in the density with linearly independent variations in χ_i and $\overline{\chi}_i$;
- Generate the form of the variations of each functional involving χ_i and $\overline{\chi_i}$: $\hat{T}$, V, J and E_{XCKE};
- Add a Lagrange multiplier to enforce the N_e-representability;
- Enforce optimality conditions of the variational principle.

The interested reader can refer to [1] for additional details on the calculation procedure.

2.3 Concluding Remarks on the Quantum Scale

After this brief analysis of the quantum scale, we reach the following conclusions:

- Schrödinger formalism represents the finest contemporary level of description. In the formalism introduced here, there is no mention in the Hamiltonian of spin-dependent magnetic interactions. These effects, as well as the relativistic ones, taken into account in Dirac's equation, are neglected. The consideration of very heavy nuclei requires the introduction of such relativistic effects.
- The wavefunction involved in the Schrödinger equation is spatially continuous, and its evolution is governed by a PDE.
- The Schrödinger equation is defined in a multidimensional space leading to the curse of dimensionality issues. It has been solved exactly for systems containing a reduced number of electrons.
- The ab-initio approximations, density functional and Hartree–Fock theories just summarized seem sometimes to be crude, but they are the only valuable route for addressing multi-electronic systems.
- The solution of the Schrödinger equation could provide an excellent description of the world at the nanometric scale, as well as accurate interatomic potentials to be used in molecular dynamics simulations.
- There are some quantum systems in which the solution explores the whole multi-dimensional configuration space, and thus remain almost intractable despite all the possible advances in the computational performances.

References

1. D.B. Cook, *Handbook of Computational Chemistry* (Oxford University Press, Oxford, 1998)
2. A. Ammar, F. Chinesta, P. Joyot, The nanometric and micrometric scales of the structure and mechanics of materials revisited: An introduction to the challenges of fully deterministic numerical descriptions. Int. J. Multiscale Comput. Eng. **6**(3), 191–213 (2008)
3. E. Cancès, M. Defranceschi, W. Kutzelnigg, C. Le Bris, Y. Maday, Computational quantum chemistry: a primer. Handb. Numer. Anal. **10**, 3–270 (2003)
4. C. Le Bris (ed.), *Handbook of Numerical Analysis, Computational Chemistry* (Elsevier, New York, 2003)

Chapter 3
Coarse-Grained Descriptions

Abstract Coarse-grained approaches are widely considered for analyzing multiatomic systems. They are based on the used of simplified interatomic potentials, that allow deriving most of the macroscopic thermomechanical properties of materials. Molecular dynamics can coarsened at its turn leading to dissipative particle dynamics and multi-particle collision dynamics. Finally, for addressing larger systems, Langevin and diffusion equations are usually considered; the last in very close connection with Brownian mechanics and its fractional variant. This chapter revisits all these physical descriptions.

Keywords Molecular dynamics · Dissipative Particle Dynamics · Multi-Particle Collision Dynamics · Langevin equation · Diffusion equation · Fractional diffusion

3.1 Molecular Dynamics

Molecular dynamics is considered nowadays to be a powerful numerical technique valuable for exploring the behavior and structure of matter at the atomic scale. This technique is becoming more and more used in different fields: computational physics and chemistry, applied mechanics and engineering, etc. Molecular dynamics is characterized by: (i) its simplicity from the conceptual point of view and (ii) the impressive CPU time required to perform realistic simulations, as well as the overly reduced physical systems that can be analyzed nowadays. In any case, this technique seems to be, despite its inherent computational cost, an excellent tool for extracting the main behavior of multi-scale models and analyzing localized phenomena that are coupled with the macroscopic scale via the use of appropriate multi-scale techniques or by applying adequate bridges between different zones analyzed at different scales.

When molecular dynamics is applied to a system in equilibrium, it allows us to determine the temporal evolution of positions, velocities and forces that, using the concepts of statistical mechanics, leads to the calculation of macroscopic properties: elastic constants, surface energy, etc. Contrastingly, when the systems evolve off equilibrium, the evolution of these fields (positions, velocities and forces) leads

© The Author(s) 2018

F. Chinesta and E. Abisset-Chavanne, *A Journey Around the Different Scales Involved in the Description of Matter and Complex Systems*, SpringerBriefs in Applied Sciences and Technology, https://doi.org/10.1007/978-3-319-70001-4_3

to the calculation of transport properties: thermal conductivity, viscosity, diffusion coefficient, defaults propagation, etc.

In what follows, we summarize the main concepts related to the essence and use of this simulation technique, emphasizing its primary difficulties, which constitute the main research domains of specialists in applied physics, mechanics and mathematics.

The heart of molecular dynamics lies in taking a population of nuclei (without considering the electrons explicitly) whose initial positions and velocities are known or simply assumed. Now, if the inter-atomic potential is known (see Sect. 1.9), its gradient gives the force applied to each nucleus due to the other nuclei, as well as to the electronic distribution defined in the whole 3D space (obviously this distribution vanishes far from the region where the nuclei are located). The first trouble arises due to the difficulty (and sometimes impossibility) of solving the Schrödinger equation to determine the wavefunction, allowing the calculation of that interatomic potential according to the procedure described in Sect. 1.9. For this reason, different potentials have been proposed and used, some of which are purely phenomenological, while others are quantum-inspired. Moreover, when one is dealing with molecules involving several kinds of atoms, there are numerous interatomic potentials to be considered, some of which are related to the strong bonds and others to the weak interactions associated with the Van der Waals effects.

3.1.1 Some Simple Examples of Pair-Wise Interatomic Potentials

Hereafter, we introduce some widely used pairwise potentials. There are many others, some of them concerning three-body and N-body potentials, that can be found in specialized books and papers.

- The hard sphere potential is defined from

$$V(\mathbf{x}; \mathbf{X}) = \begin{cases} \infty \; if \; ||\mathbf{x} - \mathbf{X}|| \leq \rho \\ 0 \quad if \; ||\mathbf{x} - \mathbf{X}|| > \rho \end{cases}, \tag{3.1}$$

  where $\mathbf{X}$ denotes the nucleus position, $\mathbf{x} \in \mathbb{R}^3$ and ρ the rigid sphere radius. This kind of potential is only applicable for deriving qualitative behaviors.
- The soft sphere potential is defined by

$$V(\mathbf{x}; \mathbf{X}) = \varepsilon \left(\frac{\sigma}{||\mathbf{x} - \mathbf{X}||} \right)^n, \tag{3.2}$$

  where ε, σ and n are now the three coefficients defining the potential. This potential only takes into account the repulsion effects, limiting its applicability to some situations.

- The square-well potential is defined by

$$V = (\mathbf{x}; \mathbf{X}) = \begin{cases} \infty \ if \ ||\mathbf{x} - \mathbf{X}|| \leq \sigma \\ -\varepsilon \ if \ \sigma < ||\mathbf{x} - \mathbf{X}|| \leq \lambda\sigma \\ 0 \quad if \ ||\mathbf{x} - \mathbf{X}|| > \lambda\sigma \end{cases}, \tag{3.3}$$

  with $\lambda > 1$, that is, the simplest potential taking into account both repulsive and attractive effects. It has been widely used for analyzing fluid properties. There are numerous variations of this potential, the Morse potential being one of the most popular:

$$V(\mathbf{x}; \mathbf{X}) = De^{-2\alpha(||\mathbf{x}-\mathbf{X}||-r_0)} - 2De^{-\alpha(||\mathbf{x}-\mathbf{X}||-r_0)}, \tag{3.4}$$

  where the coefficients D, α and r_0 depend on the considered material.
- The Lennard-Jones potential (LJ) is one of the most widely used in MD simulations. It allows us to consider both attraction and repulsion effects. Among the vast family of LJ models, perhaps the most popular is given by

$$V(\mathbf{x}; \mathbf{X}) = 4\varepsilon \left[\left(\frac{\sigma}{||\mathbf{x} - \mathbf{X}||} \right)^{12} - \left(\frac{\sigma}{||\mathbf{x} - \mathbf{X}||} \right)^{6} \right]. \tag{3.5}$$

3.1.2 *Integration Procedure*

Now, if the force acting on each nucleus is assumed to be known $\mathbf{F}_i$, $i = 1, \ldots, N_n$, which, as mentioned above, is obtained from the gradient of the considered potential, the Newton equation allows us to compute the accelerations

$$\mathbf{a}_i = \frac{\mathbf{F}_i}{m_i}, i = 1, \ldots, N_n, \tag{3.6}$$

whose integration allows us to update velocities and positions. One of the most widely used integration procedures is Verlet's algorithm. It is derived from the expansions

$$\begin{cases} \mathbf{x}(t + \Delta t) = \mathbf{x}(t) + \displaystyle\sum_{k=1}^{\infty} \frac{1}{k!} \left. \frac{d^k \mathbf{x}}{dt^k} \right|_t (\Delta t)^k \\ \mathbf{x}(t - \Delta t) = \mathbf{x}(t) + \displaystyle\sum_{k=1}^{\infty} (-1)^k \frac{1}{k!} \left. \frac{d^k \mathbf{x}}{dt^k} \right|_t (\Delta t)^k \end{cases}, \tag{3.7}$$

whose sum results in

$$\mathbf{x}(t+\Delta t)+\mathbf{x}(t-\Delta t)=2\mathbf{x}(t)+\left.\frac{d^2\mathbf{x}}{dt^2}\right|_t(\Delta t)^2+\Theta(\Delta t)^4, \tag{3.8}$$

where $\left.\frac{d^2\mathbf{x}}{dt^2}\right|_t=\mathbf{a}(t)$. This integration scheme, of fourth order, allows us to update the nuclei position without using the nuclei velocities. However, these velocities are needed in order to compute the kinetic energy, the velocity distribution, etc. To obtain an expression for computing the nuclei velocities, we proceed by subtracting both expansions

$$\mathbf{x}(t+\Delta t)-\mathbf{x}(t-\Delta t)=2\left.\frac{d\mathbf{x}}{dt}\right|_t\Delta t+\Theta(\Delta t)^3, \tag{3.9}$$

where $\left.\frac{d\mathbf{x}}{dt}\right|_t=\mathbf{v}(t)$.

There are numerous integration schemes with different properties of stability, accuracy (energy conservation), simplicity of implementation, computing time and computer memory needs, etc. In general, the integration strategies are explicit, making possible the use of massive parallel computing platforms.

3.1.3 Discussion

After this brief introduction, one could think: where are the difficulties?

Molecular dynamics simulations, despite their conceptual simplicity, are rife with diverse difficulties of different natures:

- The first and most important comes, as previously indicated, from the impossibility of using an "exact" interaction potential derived from the quantum scale. This situation is particularly delicate when we are dealing with such irregular nuclei distributions as the ones encountered in the neighborhood of defaults in crystals (dislocations, crack-tip, etc.), interfaces between different materials or in zones where different kinds of nuclei coexist.
- The second one comes from the units involved in this kind of simulation: the nuclei displacements are in the nanometric scale, the energies are on the order of the electron-volt ($1.6\ 10^{-16}$J), and the time steps are on the order of the picosecond (10^{-15}s). Thus, because of the limits in the computer's precision, a change of units is required, which can be easily performed. Moreover, some interatomic potentials in the literature are related to a particular choice of units.
- In molecular dynamics, the behavior of atoms and molecules is described in the framework of classical mechanics. Thus, the particles' energy variations are continuous. The applicability of MD depends on the validity of this fundamental hypothesis. When we consider crystals at low temperature, the quantum effects (implying discontinuous energy variations) are preponderant, and as a consequence the matter properties at these temperatures cannot be determined by MD simulations. The use of MD is restricted to temperatures higher than Debye's temperature. This

analysis is in contrast to the vast majority of MD simulations carried out nowadays. In fact, the higher the temperature (kinetic energy), the higher the velocity of particles, requiring shorter time steps in order to ensure the stability of the integration scheme. For this reason, most of the MD simulations in solid mechanics nowadays are carried out at zero degrees Kelvin or at very low temperatures but, as was just pointed out, at these temperatures, the validity of the computed MD solutions are polluted by the non-negligible quantum effects, and it is important to note that many engineering problems imply high temperatures.

- The prescription of boundary conditions is another delicate task. If the analysis is restricted to systems with free boundary conditions, then the MD simulation can be carried out without any particular treatment. In the other case, we must consider a system large enough to ensure that in the analyzed region, the impact of the free surfaces can be neglected. Another possibility lies in the prescription of periodic boundary conditions, where an atom leaving the system, for example, through the right boundary is re-injected into the domain through the left boundary. The particles located in the neighborhood of a boundary are influenced by the ones located in the neighborhood of the opposite boundary. The imposition of other boundary conditions is more delicate from both the numerical and conceptual points of view. For example, what is the meaning of prescribing a displacement on a boundary when the system is not at zero degrees? Each situation requires a deep analysis in order to define the best (the most physical) way to prescribe the boundary conditions.
- There are other difficulties related to the transient analysis. Consider a thermal system in equilibrium (in which the distribution of velocities is in agreement with the Maxwell-Boltzmann distribution). Now, we proceed to heat the system. One possibility lies in suddenly increasing the kinetic energy of each particle. Obviously, even if the resulting velocities define a physical distribution, the system remains off equilibrium because the partition between kinetic and potential energies is not the appropriate one. For this reason, we must proceed to relax the system that evolves from this initial state to that of equilibrium. Other (more physical) possibility lies in incorporating a large enough ambient region around the analyzed system (the so-called thermostat), whose particles are initially in equilibrium at the highest temperature. Now, both regions (the system and the ambient) interact, and the system initiates its heating process which reaches its equilibrium some time latter. The final state of both evolutions is the same, but the time required to reach it depends on the technique used to induce the heating. The first transient is purely numerical, whereas the second one is more physical, allowing of the identification of some transport coefficients (e.g., thermal conductivity).
- Finally, the CPU time continues to be the main limitation of MD simulations. The strongest handicap is related to the necessity of considering, at each time step and for each particle, the influence of all the other particles. Thus, the integration method seems to scale with the square of the number of particles. Even if some time can be saved in computing the neighbors search, the extremely small time steps and the extremely large number of particles required to describe real scenarios, considerably limit the range of applicability of this kind of simulations, which

is nowadays accepted as being on the order of some cubic micrometers, even when the systems are considered at very low, and thus non-physical, temperatures (close to zero °K). Despite the impressive advances in computational capabilities, high performance computing and the use of massive parallel computing platforms, the state-of-the-art does not allow for the treatment of macroscopic systems encountered in practical applications of physics, chemistry and engineering.

In conclusion, MD is conceptually very easy to understand, and not too difficult to implement, but extremely expensive in terms of computing time and resources. Its discrete nature implies simplicity, but at the same time, the technique becomes too computationally expensive to be efficiently applied to the analysis of large macroscopic systems.

3.1.4 Recovering Macroscopic Behaviors

In this section, we illustrate the emergence of macroscopic behaviors from the extremely detailed atomic description, justifying the interest in conducting molecular dynamics simulations despite the overly small scales (in space and time) attainable nowadays.

When solving the Schrödinger equation for a simple atom, example, a Lithium atom, composed of three electrons, the ground-state defines a problem in 9 dimensions, and we obtain, in agreement with the Pauli exclusion principle, the electronic distribution illustrated in Fig. 3.1. The first two electrons occupy the so-called s-orbital, an almost spherical shell around the atom nucleus, while the third one is distributed in one of the three available p-orbitals (the shapes of these orbitals are depicted in Fig. 3.2).

The same procedure can be applied to more complex systems composed of different atoms in order to investigate the electronic distribution in simple molecules. As discussed in the previous chapter, when considering the Schrödinger equation, in general, we assume the nuclei positions to be fixed and compute the associated electronic distribution, which consequently will depend parametrically on the nuclei positions. The electronic distribution and, in particular, its ground state are related

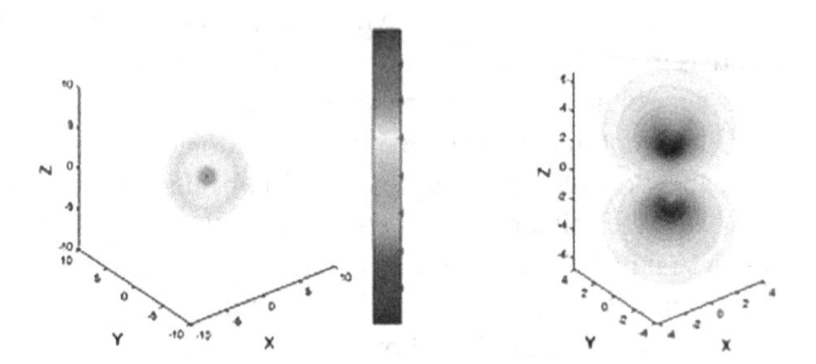

Fig. 3.1 Lithium atom: electronic distribution (color figure online)

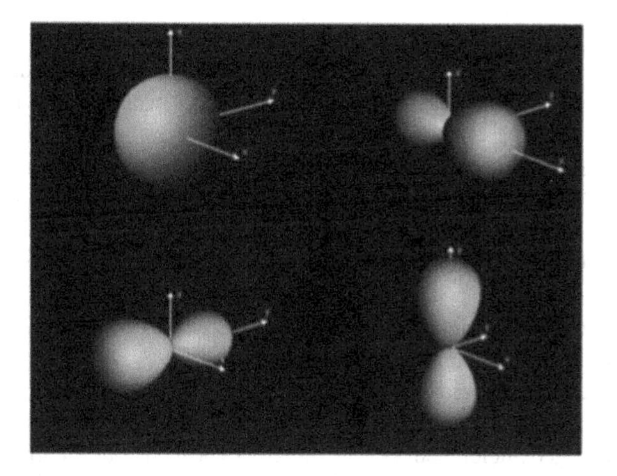

Fig. 3.2 Shape of the s and p electronic orbitals (color figure online)

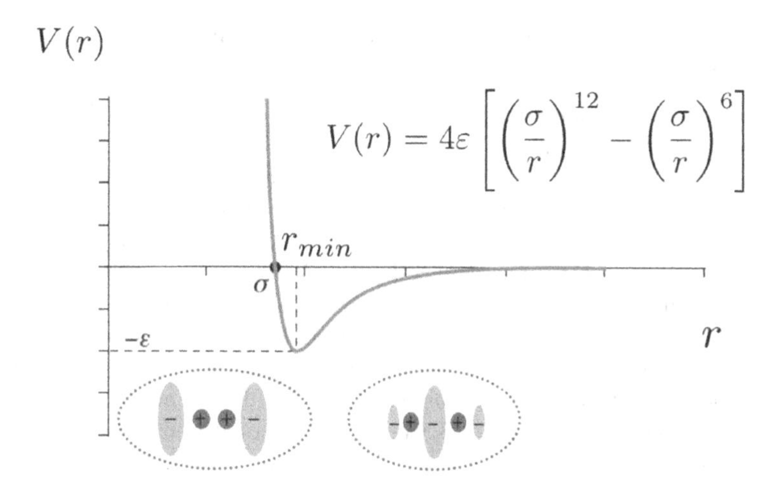

Fig. 3.3 Two-particles potential $V(r)$ evolution with the inter-particle distance r (color figure online)

to the lowest energy, which as just mentioned, will depend on the chosen location of the nuclei.

When considering a simple molecule composed of two atoms, we can compute the energy associated with the ground state as a function of the relative distance between both nuclei. By assuming both atoms to be on the x-coordinate axis, the first located at X (e.g. $X = 0$) and the second one placed at $X + r$ (r when $X = 0$), with $r \in \mathbb{R}^+$, the ground-state energy (the so-called two-particles potential) will depend on the inter-nuclei distance r, $V(r)$, and as discussed above, it is, in general, adequately represented by the Lennard-Jones potential. A typical representation is depicted in Fig. 3.3, in which different facts can be noticed:

- When both particles move apart, $r \rightarrow \infty$, both the potential and its gradient vanish. Since the inter-particle force acts in the opposite direction as that of potential gradient, being $\nabla V(r)|_{r\rightarrow\infty} \approx 0$, the resulting force vanishes and both particles become free.
- When particles approach one another the solution of the Schrödinger equation results in an electronic distribution that concentrates in the region between both nuclei. Thus, electrons become a sort of glue that compensate for the mutual repulsion of both positively charged nuclei, ensuring the molecule's stability. The potential gradient for $r > r_{min}$ is positive and the resulting force negative. The attractive force tends to approach the test nucleus to the one assumed to be located at the origin $X = 0$ and kept at rest.
- When the nuclei are too close to one another, the electronic distribution that results from the Schrödinger solution leaves the region in between both nuclei, and both positively charged nuclei then become directly exposed to one another. A repulsion appears (negative potential gradient, then positive force) that tends to separate both nuclei.
- The equilibrium distance (r_{min} in Fig. 3.3), the one for which the potential gradient vanishes, defines the reference interatomic distance.

When the molecule only contains the energy involved in the potential (blue curve), no remaining part exists that is related to the kinetic energy. However, if an extra-energy ΔE is communicated to the system, as illustrated in Fig. 3.4, the energy gap $E - V(r)$ represents the kinetic energy available in the molecule. Thus, if we assume the nucleus located at $X = 0$ to be at rest, the one initially located at r_{min} must move in the permitted region, the one defined by $E - V(r) \geq 0$ (red arrow in Fig. 3.4). Obviously, in the periodic movement, the two limits of the interval are defined by $E - V(r) = 0$ and the kinetic energy vanishes at both points. Contrastingly when $r = r_{min}$, the kinetic energy $E - V(r_{min})$ reaches its highest value. It is a simple result of the fact that atoms vibrate continuously when there is an excess of energy.

Another important fact that Fig. 3.3 reflects is the dilatation properties. As the potential curve around its minimum at $r = r_{min}$ is non-symmetric, it implies that,

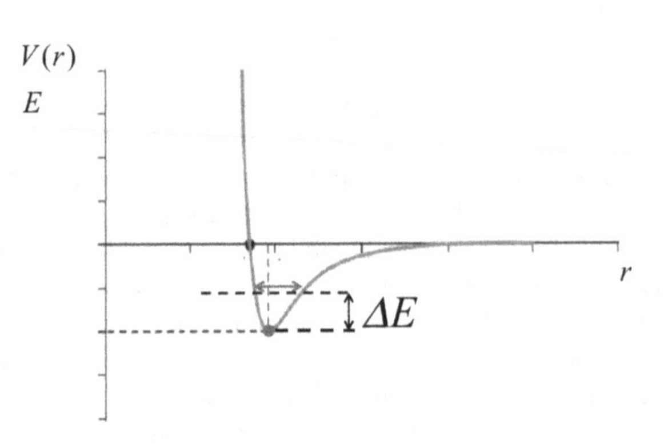

Fig. 3.4 Communicating kinetic energy to a molecular system (color figure online)

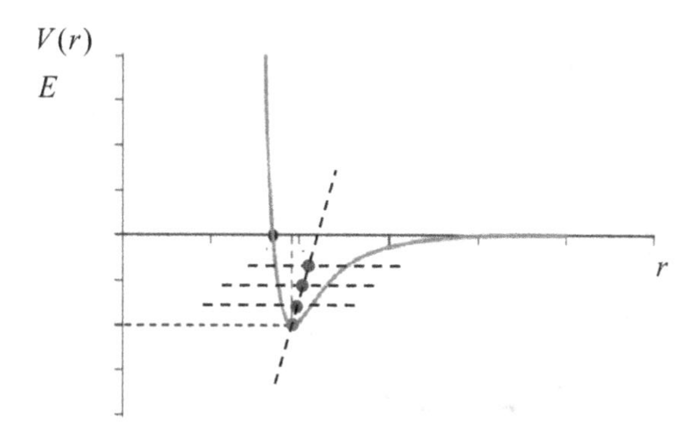

Fig. 3.5 Explaining dilatation in materials (color figure online)

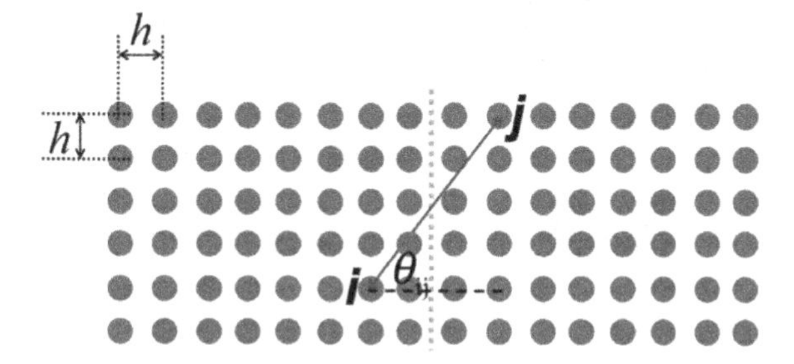

Fig. 3.6 Forces applied between pairs of nuclei on each side of the considered surface are calculated and then projected (color figure online)

during heating, the energy level E varies, and with it the vibration amplitude, as illustrated in Fig. 3.5. Thus, the average position $\overline{r} = \frac{1}{T} \int_0^T r(t) dt$ (with T as the vibration period) increases as the energy gap increases, explaining standard dilatation in materials.

Finally, because the potential gradient results in interatomic forces, considering a virtual surface within a solid composed of many nuclei and calculating the resultant of all the inter-particles forces crossing that surface, forces that depend on the relative distance between each pair of nuclei, by projecting on the different coordinate axes and dividing by the surface, allows us to define the Cauchy stress, as illustrated in Fig. 3.6. Moreover, if the potential $V(r)$ around its minimum at r_{min} is parabolic, the related force becomes linear, with its slope defining the elastic modulus.

On the other hand, the maximum force that a bond can resists is the maximum potential gradient in the region $r \geq r_{min}$.

Finally, we describe the emergence of transport properties. For that purpose, we consider a circular ring composed of two regions, one at high temperature and the other (much larger) at lower temperature, as illustrated in Fig. 3.7 (left). At the initial time, we assume Maxwellian velocity distributions in both regions, as depicted in Fig. 3.7 (right).

The Verlet integration scheme is applied, and as expected, because the atoms in the coldest region close to the warmest one are pushed with higher intensity from the heated side, they increase their velocity (temperature), while the ones in the warmest

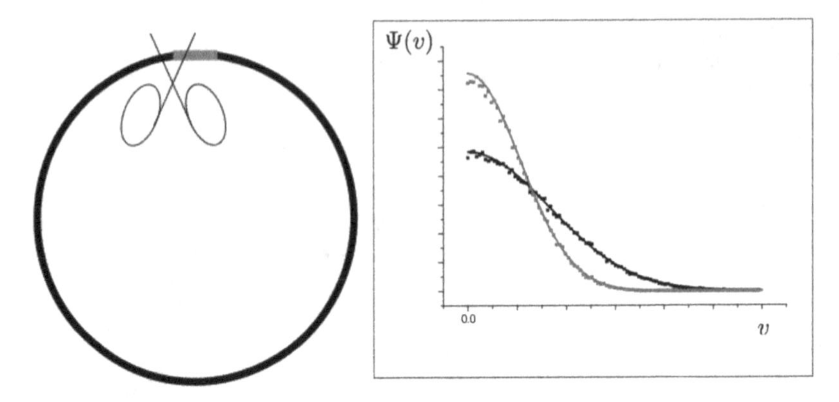

Fig. 3.7 Unbounded circular ring composed of two regions (left) with initial Maxwellian velocity distributions (right). The scissors indicates that the domain will be represented unfolded in Fig. 3.8 (color figure online)

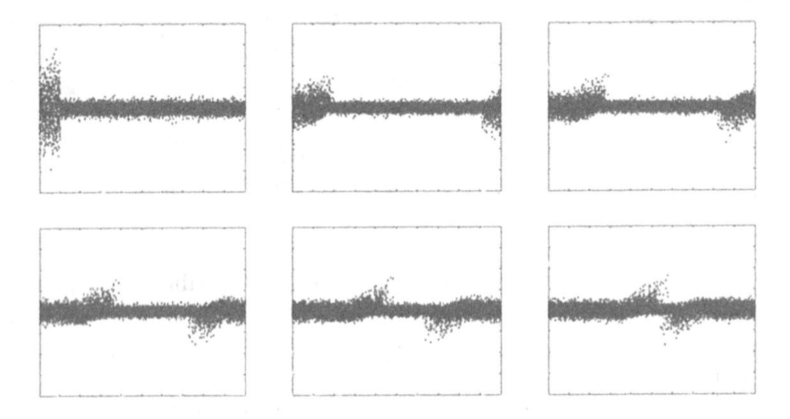

Fig. 3.8 Heat propagation by conduction in the ring depicted in Fig. 3.7 using an unfolded representation. Thermal diffusivity emerges from molecular dynamics simulations (color figure online)

region pushing them, lose energy, as depicted in Fig. 3.8. Thus, the heat propagates from the warmest region to the coldest one, and molecular dynamics again allows us to explain the macroscopic Fourier law. The speed of propagation determines the material thermal diffusivity.

3.1.5 Molecular Dynamics-Continuum Mechanics Bridging Techniques

The above-mentioned difficulties in fully performing molecular dynamics simulations motivated the proposal of hybrid techniques that apply MD in the regions where the unknown field varies in a non-uniform way (molecular dynamics model) and a standard finite element approximation in those regions where the unknown field variation can be considered to be uniform (continuous model). The main questions surrounding these bridging strategies concern: (i) the kinematics representations in both models; (ii) the transfer conditions on the MD and continuous models interface; and (iii) the macroscopic constitutive equation to be employed in the continuous model.

Different alternatives exist, and the construction of such bridges is one of the most actives topics in computational mechanics nowadays. The spurious reflection of the high frequency parts of waves is one of the main issues. We would like simply to mention two families of bridging techniques, giving some key references:

1. The quasi-continuum method proposed by Tadmor and Ortiz can be applied for establishing bridges between MD and continuum models [1]. It links atomistic and continuum models through the use of a finite element to reduce the full set of atomistic degrees of freedom. Thus, in the regions where the solution evolution is non-uniform, the full atomistic description is retained. In those regions where the solution evolves uniformly, it is possible to select a (reduced) number of representative atoms to describe the kinematics (via the finite element interpolation) and the energy of the body. This observation suggests a division of the representative atoms into two classes: (i) nonlocal atoms whose energies are computed by an explicit consideration of all of their neighbors according to the MD practice; and (ii) local atoms whose energies are computed from the local deformation gradients using the Cauchy-Born (which states that the atoms in the deformed crystal will move to positions dictated by the existing gradients of displacements). Thus, instead of using the phenomenological constitutive equation in the continuum model, such as Hooke's law, this technique uses atomistic calculations to inform the energetic statement of the continuum mechanics variational principle. Special treatments must be employed to avoid the "ghost" forces appearing in the transition zones.
2. Ben Dhia proposed, in 1998 [2], a superposition technique able to define efficient and accurate bridges in multiscale, multimodel and multiphysics (the Arlequin method). These ideas have recently been adopted by numerous teams who have emphasized, in particular, the excellent potential of this approach for coupling MD and continuum models [3, 4]. The main idea is to define an overlapping region in which models (MD and continuum) coexist. The energy in the fully MD and continuum regions is defined without ambiguity, however, the Arlequin approach defines the energy in the overlapping region from a linear combination of the ones related to the MD and continuum models. At a certain point, the weights of both energies depend on the position of that point. When the point (inside the

overlapping region) approaches the fully MD region, the weight related to the MD energy approaches one and the weight related to the continuum energy vanishes. When the point approaches the fully continuum domain, the inverse tendencies are found, and when the point is located inside the overlapping region, the weights are computed according to the distance to each region under the constraint of unit sum of both. Thus, the energy can be perfectly defined everywhere. The kinematics are also defined without ambiguity on the MD and continuum regions, and a weak equality of the MD and continuum displacements is enforced in the overlapping region.

3.1.6 Coarse-Grained Molecular Dynamics: DPD and MPCD

Conventional MD provides too much detail of the actual motion of the molecules of a fluid. If one is interested in hydrodynamic behavior, one can look at a more coarse-grained level.

3.1.6.1 Dissipative Particle Dynamics – DPD

The particles in DPD are not real molecules, but rather a sort of molecule cluster called a fluid particle. By introducing dissipation into a molecular-dynamics simulation, one expects to observe hydrodynamic behavior with a considerably smaller number of particles, thus reducing the computational effort. In the DPD, not only the number of particles is conserved, but also the total momentum of the system.

For a fluid particle with momentum $\mathbf{p}_i$, its time derivative is equal to the net force applied to it

$$\dot{\mathbf{p}}_i = \sum_{j \neq i} \left(\mathbf{F}_{ij}^C + \mathbf{F}_{ij}^D + \mathbf{F}_{ij}^R \right), \tag{3.10}$$

where $\bullet^C$, $\bullet^D$ and $\bullet^R$ refer to the conservative, dissipative and random forces, respectively. Because of the fluctuation-dissipation theorem, the last two forces must apply jointly.

The conservative force comes from the gradient of a potential (as it was the case in molecular dynamics). Galilean invariance requires that both the dissipative and random forces depend on the relative positions and velocities, $\mathbf{r}_{ij} = \mathbf{r}_i - \mathbf{r}_j$ and $\mathbf{v}_{ij} = \mathbf{v}_i - \mathbf{v}_j$, respectively, where $\mathbf{r}_i$ and $\mathbf{v}_i$ are the position and velocity of particle i.

The simplest form of these forces guaranteeing these hypotheses is

$$\mathbf{F}_{ij}^D = -\gamma \omega_D(r_{ij})(\mathbf{e}_{ij} \cdot \mathbf{v}_{ij})\mathbf{e}_{ij}, \tag{3.11}$$

and

$$\mathbf{F}_{ij}^{R} = \sigma \omega_R(r_{ij})\, \zeta_{ij} \mathbf{e}_{ij}, \tag{3.12}$$

where $r_{ij} = \|\mathbf{r}_i - \mathbf{r}_j\|$, $\mathbf{e}_{ij} = \frac{\mathbf{r}_i - \mathbf{r}_j}{r_{ij}}$ and ζ_{ij} are Gaussian white-noise random variables such that $\zeta_{ij} = \zeta_{ji}$ for ensuring the momentum conservation, with the stochastic properties

$$\begin{cases} \langle \boldsymbol{\zeta}_{ij}(t) \rangle = 0 \\ \langle \boldsymbol{\zeta}_{ij}(t)\, \boldsymbol{\zeta}_{kl}(t') \rangle = \left(\delta_{ik}\delta_{jl} + \delta_{il}\delta_{jk}\right) \delta(t - t') \end{cases}, \tag{3.13}$$

and where ω_D and ω_R define the interaction horizon for dissipation and random forces.

Even if many simulations considered that $\omega_D = \omega_R$, in [5], it was proved that the verification of the fluctuation-dissipation theorem requires one to consider that $\omega_R(r) = \sqrt{\omega_D(r)}$.

3.1.6.2 Multi-Particle Collision Dynamics – MPCD

MPCD consists of two stages, streaming and collision, ensuring the conservation of particles, mass and linear momentum, as well as a sufficient degree of isotropy to reproduce hydrodynamic behavior [6]. Its computational simplicity and the possibility of considering highly parallel implementations explains the growing interest within the scientific community.

The system is composed of N particles, each of them located at position $\mathbf{r}_i(t)$ and having a velocity $\mathbf{v}_i(t)$. We assume that collisions only occur at discrete time-intervals τ. Thus, the new position of particle i at time $t + \tau$ reads as

$$\mathbf{r}_i^*(t + \tau) = \mathbf{r}_i(t) + \mathbf{v}_i \tau. \tag{3.14}$$

Now, for the purpose of considering collisions, we proceed as follows. The volume V is divided into N_c cells, $\mathscr{C}_\alpha$, $\alpha = 1, \cdots, N_c$, each of them instantaneously containing N_α particles. We can define the α-cell center of mass velocity $\mathbf{V}_\alpha$ from

$$\mathbf{V}_\alpha = \frac{1}{N_\alpha} \sum_{i \in \mathscr{C}_\alpha} \mathbf{v}_i. \tag{3.15}$$

Then, a random rotation is assigned to each cell $\boldsymbol{\Omega}_\alpha$ and is applied to all the particles that it contains, which results in the post-collision velocities given by

$$\mathbf{v}_i^* = \mathbf{V}_\alpha + \boldsymbol{\Omega}_\alpha(\mathbf{v}_i - \mathbf{V}_\alpha), \quad \forall i \in \mathscr{C}_\alpha. \tag{3.16}$$

MPCD has the following properties:

- Mass conservation is trivially ensured;
- Linear momentum is conserved at each cell. The proof proceeds as follows:

$$\sum_{i\in\mathscr{C}_\alpha} m\mathbf{v}_i^* = \sum_{i\in\mathscr{C}_\alpha} m\left(\mathbf{V}_\alpha + \boldsymbol{\Omega}_\alpha(\mathbf{v}_i - \mathbf{V}_\alpha)\right), \tag{3.17}$$

since the rotation is the same for all of the particles into the cell $\mathscr{C}_\alpha$

$$\sum_{i\in\mathscr{C}_\alpha} m\boldsymbol{\Omega}_\alpha(\mathbf{v}_i - \mathbf{V}_\alpha) = m\boldsymbol{\Omega}_\alpha \sum_{i\in\mathscr{C}_\alpha} (\mathbf{v}_i - \mathbf{V}_\alpha)$$

$$= m\boldsymbol{\Omega}_\alpha \left(\left(\sum_{i\in\mathscr{C}_\alpha} \mathbf{v}_i \right) - N_\alpha \mathbf{V}_\alpha \right) = 0, \tag{3.18}$$

it finally results that

$$\sum_{i\in\mathscr{C}_\alpha} m\mathbf{v}_i^* = \sum_{i\in\mathscr{C}_\alpha} m\mathbf{V}_\alpha = \sum_{i\in\mathscr{C}_\alpha} m\mathbf{v}_i. \tag{3.19}$$

- Finally, energy conservation follows from

$$\sum_{i\in\mathscr{C}_\alpha} m\left(\mathbf{v}_i^*\right)^2 = \sum_{i\in\mathscr{C}_\alpha} m\left(\mathbf{V}_\alpha + \boldsymbol{\Omega}_\alpha(\mathbf{v}_i - \mathbf{V}_\alpha)\right)^2. \tag{3.20}$$

By developing the squared term

$$\begin{aligned}(\mathbf{V}_\alpha + \boldsymbol{\Omega}_\alpha(\mathbf{v}_i - \mathbf{V}_\alpha))^2 &= \left(\mathbf{V}_\alpha^T + (\mathbf{v}_i - \mathbf{V}_\alpha)^T \boldsymbol{\Omega}_\alpha^T\right)(\mathbf{V}_\alpha + \boldsymbol{\Omega}_\alpha(\mathbf{v}_i - \mathbf{V}_\alpha)) \\ &= \mathbf{V}_\alpha^T\mathbf{V}_\alpha + (\mathbf{v}_i - \mathbf{V}_\alpha)^T \boldsymbol{\Omega}_\alpha^T \boldsymbol{\Omega}_\alpha(\mathbf{v}_i - \mathbf{V}_\alpha) + \\ & (\mathbf{v}_i - \mathbf{V}_\alpha)^T \boldsymbol{\Omega}_\alpha^T \mathbf{V}_\alpha + \mathbf{V}_\alpha^T \boldsymbol{\Omega}_\alpha(\mathbf{v}_i - \mathbf{V}_\alpha),\end{aligned} \tag{3.21}$$

which taking into account the orthogonality of $\boldsymbol{\Omega}_\alpha$, $\boldsymbol{\Omega}_\alpha^T\boldsymbol{\Omega}_\alpha = \mathbf{I}$, leads to

$$\begin{aligned}&(\mathbf{V}_\alpha + \boldsymbol{\Omega}_\alpha(\mathbf{v}_i - \mathbf{V}_\alpha))^2 = \\ &\mathbf{V}_\alpha^T\mathbf{V}_\alpha + (\mathbf{v}_i - \mathbf{V}_\alpha)^T(\mathbf{v}_i - \mathbf{V}_\alpha) + (\mathbf{v}_i - \mathbf{V}_\alpha)^T \boldsymbol{\Omega}_\alpha^T \mathbf{V}_\alpha + \mathbf{V}_\alpha^T \boldsymbol{\Omega}_\alpha(\mathbf{v}_i - \mathbf{V}_\alpha).\end{aligned} \tag{3.22}$$

Now, when applying the sum, and taking into account (3.18), the last two terms in Eq. (3.22) vanish, resulting in

$$\sum_{i\in\mathscr{C}_\alpha} m\left(\mathbf{v}_i^*\right)^2 = \sum_{i\in\mathscr{C}_\alpha} m\left(\mathbf{V}_\alpha^T\mathbf{V}_\alpha + \mathbf{v}_i^T\mathbf{v}_i - \mathbf{V}_\alpha^T\mathbf{v}_i - \mathbf{v}_i^T\mathbf{V}_\alpha + \mathbf{V}_\alpha^T\mathbf{V}_\alpha\right), \tag{3.23}$$

which taking into account

$$\sum_{i\in\mathscr{C}_\alpha} \mathbf{V}_\alpha^T\mathbf{v}_i = N_\alpha \mathbf{V}_\alpha^T\mathbf{V}_\alpha, \tag{3.24}$$

and

$$\sum_{i \in \mathscr{C}_\alpha} \mathbf{V}_\alpha^T \mathbf{V}_\alpha = N_\alpha \mathbf{V}_\alpha^T \mathbf{V}_\alpha, \tag{3.25}$$

ultimately results in

$$\sum_{i \in \mathscr{C}_\alpha} m \left(\mathbf{v}_i^*\right)^2 = \sum_{i \in \mathscr{C}_\alpha} m \mathbf{v}_i^2. \tag{3.26}$$

When different type of particle are involved, each one is considered with its mass. When coupling fluid and rigid or deformable objects, it suffices to use the same collision operator; the streaming must be applied to the fluid particles, and the kinematics of the particles composing the object will ultimately be derived from the Hamiltonian defining the object's behavior.

If one uses a fixed grid to define the collision cells, one breaks the system symmetry under Galilei transformations. Imagine a fluid in a situation in which the mean free path during the time-step is substantially smaller than the collision cell size Δx. Then, given a particular collision cell, the set of particles in that cell at time t is going to contain mostly the same members as at the next time step (collision time). Statistically, the states of those particles are therefore going to be correlated over a significant amount of time compared to the streaming time-step.

However, if one superimposes a global, fixed and non-zero velocity on the entire system, the correlation time changes in general, since now, the sets of particles in the cell at different times may share fewer members. This means that the statistical properties of a system depend on the observer's inertial frame, breaking Galilean symmetry.

This deficiency of broken Galilean symmetry can be eliminated by independently sampling three random numbers defining the vector $\mathbf{S}$ and shifting either the entire collision cell grid by $\mathbf{S}$ with reference to its fixed position in the previous scenario, or, equivalently, by shifting the positions of all MPC particles by $-\mathbf{S}$.

3.2 Brownian Dynamics: A Step Towards Coarse-Grained Models

Looking for significant computing time savings, different coarse-grained models have been proposed and successfully used. One of these approaches is the Brownian dynamics (BD) simulation and some variants of it.

We will summarize the main ideas related to this framework by considering a simple scenario, the one revealed by Robert Brown in 1827. Robert Brown observed that small and light particles immersed in a fluid show a kind of erratic trajectory on the microscopic scale, even if there is an average movement tendency induced by gravity or the fluid drift. This fact is justified today by the fact that atoms are in constant movement (since their kinetic energy is proportional to the temperature), exploring all possible movement directions. This constitutes the foundations of the kinetic theory of gases proposed by Maxwell and Boltzmann. Due to this pseudo-erratic movement, the fluid atoms sometimes impact the suspended particles, inducing a change in their quantity of movement and then in their direction of motion.

From the previous analysis, one could expect that the only possibility for simulating this kind of scenario lies in taking into account all the atoms existing in the fluid, as well as all the small test particles immersed inside, and proceeding in the context of MD. However, the number of particles involved is too large with respect to computer availabilities nowadays. Remember that the state-of-the-art in MD simulation allows us to treat systems on the order of some cubic micrometers. For this reason, one could imagine the removal of all of the fluid atoms but with retainment of their effects on the particles of interest, that is, the impact statistics.

3.2.1 The Langevin Equation

In the situation just described, the motion equation for each one of the N particles in the suspension reads, according to classical mechanics, as

$$m\frac{d^2x_i}{dt^2} + \xi\frac{dx_i}{dt} = \Re_i(t); \quad i = 1, \ldots, N. \tag{3.27}$$

For the sake of simplicity, we assume 1D particle motions (the extension to 3D is straightforward) whose positions are defined by the coordinate x_i. In Eq. (3.27), also known as Langevin's equation, m is the particles's mass, ξ the friction coefficient (which in the case of the Stokes drag for spherical particles, reads as $\xi = 6\pi r\eta$, with r the sphere radius and η the fluid viscosity) and $\Re_i$ the impact to which particle i is exposed from the ambient atoms. In this equation, $m\frac{d^2x_i}{dt^2}$ represents the inertia term, $\xi\frac{dx_i}{dt}$ the viscous resistance and $\Re_i(t)$ the external forces originated by the atoms impacts. The viscous force is proportional to the difference between the particle velocity and the unperturbed fluid velocity at the position of the particle $v_f(x_i)$, implying, in general, a term in the form: $\xi\left(\frac{dx_i}{dt} - v_f(x_i)\right)$ that simplifies to the viscous term considered in Eq. (3.27), when the unperturbed fluid remains at rest. Equation (3.27) represents the balance of forces applied to each particle, whose integration only needs the specification of the impact forces. Obviously, this statistical distribution becomes well defined as soon as both its mean value and its variance (or the associated standard deviation) are given. In an isotropic medium, one could expect that the average of all the impacts is zero. Moreover, as these impacts are uncorrelated, we can equate time and ensemble averages, and write

$$\langle\Re(t)\rangle = \frac{1}{N}\sum_{i=1}^{N}\Re_i(t) = 0. \tag{3.28}$$

Now, if we have access to the standard deviation of the impact statistics, the motion equation (3.27) will be perfectly defined, making its numerical integration possible. For this purpose, we introduce a highly valuable result in statistics, the central limit theorem that states that the sum $Y_n(t)$ of any n random variables $y_i(t)$, converges to a normal distribution whose mean value and variance result in the sum of mean values $\langle y_i \rangle$ and variances $(\Delta y_i)^2$ of the random variables involved, i.e.,

$$Y_n(t) \equiv \sum_{i=1}^{i=n} y_i(t) \underset{n\to\infty}{\longrightarrow} \mathcal{N}\left(\sum_{i=1}^{i=n} \langle y_i \rangle, \sum_{i=1}^{i=n} (\Delta y_i)^2\right). \tag{3.29}$$

As the time elapsed between two consecutive impacts δt is much lower than the simulation time step Δt employed to integrate the motion equation (3.27), we can define the action $B_{\Delta t} \equiv \int_t^{t+\Delta t} \frac{\Re(t)}{m} dt$ that, using the fact that $\delta t \ll \Delta t$, as well as the central limit theorem, results in

$$\int_t^{t+\Delta t} \frac{\Re(t)}{m} dt = \sum_{j=1}^{p} \frac{\Re(t_j)}{m} \delta t \to \mathcal{N}(0, q\Delta t), \tag{3.30}$$

where $p = \frac{\Delta t}{\delta t} \gg 1$, justifying that the variance of the resulting normal distribution will be proportional to the time step Δt. In what follows, we will try to identify the value of factor q in Eq. (3.30).

The integration of the Langevin equation (3.27) (see [7] for details) leads to the equilibrium velocity distribution

$$W(v, t \to \infty) = \sqrt{\frac{\xi}{m\pi q}}\, e^{-\frac{\xi v^2}{mq}}, \tag{3.31}$$

which must coincide with the equilibrium distribution associated with the canonical ensemble (based on the equipartition theorem), implying

$$q = \frac{2\xi K_b T}{m^2}, \tag{3.32}$$

where K_b is the Boltzmann constant and T the temperature. Expression (3.32) states that the strength of the Brownian force is related to the viscous force (fluctuation-dissipation relation).

In conclusion, the simulation of the Langevin equation, or its more general expression involving an external potential $V(x)$

$$m\frac{d^2 x_i}{dt^2} + \xi\frac{dx_i}{dt} + \frac{dV}{dx} = \Re_i(t); \quad i = 1, \ldots, N, \tag{3.33}$$

only requires, for a population of particles large enough, the use of an appropriate numerical integration scheme and the consideration of a normal random variable to model the impacts

$$\frac{\Re_{\Delta t}}{m} = \mathcal{N}\left(0, \frac{2\xi K_b T}{m^2} \Delta t\right). \tag{3.34}$$

3.2.2 *From Diffusion to Anomalous Diffusion*

In his pioneering work, Einstein assumed the increment of the particle position Δ in the unbounded one-dimensional axis x to be a random variable, with a probability density given by $\phi(\Delta)$. The particles balance can be expressed by both

$$\rho(x, t + \tau) = \rho(x, t) + \frac{\partial \rho}{\partial t}\tau + \Theta(\tau^2), \tag{3.35}$$

and

$$\rho(x, t + \tau) = \int_{\mathbb{R}} \rho(x + \Delta, t)\phi(\Delta)d\Delta. \tag{3.36}$$

Developing $\rho(x + \Delta, t)$,

$$\rho(x + \Delta, t) = \rho(x, t) + \frac{\partial \rho}{\partial x}\Delta + \frac{1}{2}\frac{\partial^2 \rho}{\partial x^2}\Delta^2 + \Theta(\Delta^3) \tag{3.37}$$

which injected into the right-hand side of Eq. (3.35) and taking into account the normality and expected symmetry

$$\begin{cases} \int_{\mathbb{R}} \phi(\Delta)d\Delta = 1 \\ \int_{\mathbb{R}} \Delta\phi(\Delta)d\Delta = 0 \end{cases}, \tag{3.38}$$

leads, after equating Eqs. (3.35) and (3.36), to

$$\rho(x, t) + \frac{\partial \rho}{\partial t}\tau = \rho(x, t) + \frac{1}{2}\frac{\partial^2 \rho}{\partial x^2}\int_{\mathbb{R}} \Delta^2\phi(\Delta)d\Delta, \tag{3.39}$$

or

$$\frac{\partial \rho}{\partial t}\tau = \frac{1}{2}\frac{\partial^2 \rho}{\partial x^2}\int_{\mathbb{R}} \Delta^2\phi(\Delta)d\Delta. \tag{3.40}$$

Now, by defining the diffusion coefficient D from

$$D = \frac{1}{2\tau}\int_{\mathbb{R}} \Delta^2\phi(\Delta)d\Delta, \tag{3.41}$$

the particle balance, also known as the diffusion equation, is given by

$$\frac{\partial \rho}{\partial t} = D \frac{\partial^2 \rho}{\partial x^2}. \tag{3.42}$$

The integration of this equation assuming that all the particles are localized at the origin at the initial time, $\rho(x, t = 0) = \delta(x)$, leads to

$$\rho(x, t) = \frac{1}{\sqrt{4\pi D t}} e^{-\frac{x^2}{4Dt}}, \tag{3.43}$$

whose second order moment (variance) scales with the time

$$\langle x^2 \rangle = 2Dt, \tag{3.44}$$

that is, the mean squared displacement scales with the elapsed time t, and the diffusion coefficient D.

By applying the Fourier transform with respect to the space coordinate to the diffusion equation (3.42), and denoting by $\rho(k, t)$ the Fourier transform of the particle density $\rho(x, t)$, i.e., $\rho(k, t) = \mathscr{F}[\rho(x, t)]$ with k as the so-called wave number, it results that

$$\frac{\partial \rho(k, t)}{\partial t} = -Dk^2 \rho(k, t), \tag{3.45}$$

which has important physical consequences. A general function contains many frequencies, and as can be appreciated from the previous equation, the lower the wavelength (higher values of k), the faster its relaxation. Thus, the highest frequencies are expected to disappear very rapidly compared to the lowest ones. Moreover, in the case of negative diffusion (encountered in phase separation), the highest frequencies grow faster than the lowest ones, and if no mechanism limits the growing process, it diverges.

3.2.2.1 The Diffusion Equation from a Random Walk Perspective

In this section, we revisit the derivation of the diffusion equation from a random walk perspective. Again for the sake of simplicity, we restrict our discussion to the 1D case, with the x-axis equipped with a grid of size Δx. We assume that in a discrete time step Δt, the test particle is assumed to jump to one of its nearest neighbour sites, with random direction. Such a process can be modeled by the master equation that writes, at the site j,

$$W_j(t + \Delta t) = \frac{1}{2} W_{j+1}(t) + \frac{1}{2} W_{j-1}(t), \tag{3.46}$$

where $W_j(t)$ is the probability of having the particle at site j at time t and the prefactor 1/2 accounts for the direction isotropy of the jumps.

Now, the usual developments can be performed:

$$\begin{cases} W_i(t+\Delta t) = W_j(t) + \frac{\partial W_j(t)}{\partial t}\Big|_t \Delta t + \Theta(\Delta t^2) \\ W_{j+1}(t) = W_j(t) + \frac{\partial W(t)}{\partial x}\Big|_j \Delta x + \frac{1}{2}\frac{\partial^2 W(t)}{\partial x^2}\Big|_j \Delta x + \Theta(\Delta x^3) \\ W_{j-1}(t) = W_j(t) - \frac{\partial W(t)}{\partial x}\Big|_j \Delta x + \frac{1}{2}\frac{\partial^2 W(t)}{\partial x^2}\Big|_j \Delta x - \Theta(\Delta x^3) \end{cases}, \tag{3.47}$$

which injected into Eq. (3.46), yields

$$\frac{\partial W}{\partial t} = D\frac{\partial^2 W}{\partial x^2}, \tag{3.48}$$

with D defined in the limit of $\Delta x \rightarrow 0$ and $\Delta t \rightarrow 0$ by

$$D = \frac{\Delta x^2}{2\Delta t}, \tag{3.49}$$

which leads to the diffusion equation derived in the previous section.

3.2.2.2 Anomalous Diffusion and the Continuous Time Random Walk

In complex fluids, micro-rheological experiments often exhibit anomalous sub-diffusion or sticky diffusion, in which the mean square displacement of Brownian tracer particles is found to scale as $\langle x^2 \rangle \propto t^\alpha$, $0 < \alpha < 1$ (see [8] and the references therein). In these cases, the use of non-integer derivatives can constitute an appealing alternative, as it allows one to correctly reproduce the observed physical behaviour while keeping the model as simple as possible. Moreover, from a physical point of view, the use of non-integer derivatives introduces a degree of non-locality that seems to be in agreement with the intrinsic nature of the physical system.

In order to move towards anomalous diffusion, we consider continuous time random walks (CTRW) that will lead to a fractional diffusion equation in the same manner as standard random walks led to the usual diffusion equation.

Before developing that equation, we will summarize the most salient concepts related to fractional derivatives.

Fractional Derivatives

There are many books on fractional calculus and fractional differential equations (e.g., [9, 10]). We summarize here the main concepts needed to understand the developments carried out below.

We start with the formula attributed to Cauchy for evaluating the n-th integration, $n \in \mathbb{N}$, of a function $f(t)$:

$$J^n f(t) := \int \cdots \int f(\tau)\, d\tau = \frac{1}{(n-1)!} \int_0^t (t-\tau)^{n-1} f(\tau)\, d\tau. \tag{3.50}$$

This can be rewritten as

$$J^n f(t) = \frac{1}{\Gamma(n)} \int_0^t (t-\tau)^{n-1} f(\tau)\, d\tau, \tag{3.51}$$

where $\Gamma(n) = (n-1)!$ is the gamma function. The latter being, in fact, defined for any real value $\alpha \in \mathbb{R}$, we can define the fractional integral from

$$J^\alpha f(t) := \frac{1}{\Gamma(\alpha)} \int_0^t (t-\tau)^{\alpha-1} f(\tau)\, d\tau. \tag{3.52}$$

Now, if we consider the fractional derivative of order α and select an integer $m \in \mathbb{N}$ such that $m-1 < \alpha < m$, then it suffices to consider an integer m-order derivative combined with a $(m-\alpha)$ fractional integral (this is depicted in sketch form in Fig. 3.9). Obviously, we could take the derivative of the integral or the integral of the derivative, resulting in the left and right-hand definitions of the fractional derivative usually denoted by $D^\alpha f(t)$ and $D_*^\alpha f(t)$, respectively,

$$D^\alpha f(t) = \begin{cases} \frac{d^m}{dt^m}\left(\frac{1}{\Gamma(m-\alpha)} \int\limits_0^t \frac{f(t)}{(t-\tau)^{\alpha+1-m}} d\tau\right), & m-1 < \alpha < m \\ \frac{d^m f(t)}{dt^m}, & \alpha = m \end{cases}, \tag{3.53}$$

and

$$D_*^\alpha f(t) = \begin{cases} \frac{1}{\Gamma(m-\alpha)} \left(\int\limits_0^t \frac{\frac{d^m f(t)}{dt^m}}{(t-\tau)^{\alpha+1-m}} d\tau\right), & m-1 < \alpha < m \\ \frac{d^m f(t)}{dt^m}, & \alpha = m \end{cases}. \tag{3.54}$$

Because these approaches to the fractional derivative began with an expression for the repeated integration of a function, one could consider a similar approach for the derivative. This was the route considered by Grunwald and Letnikov – GL – defining the so-called 'differintegral' that leads to the fractional counterpart of the usual finite differences.

The Fourier transform of a fractional derivative of order α reads as $\mathscr{F}(g(t);\omega) = (i\omega)^\alpha \mathscr{G}(\omega)$, and analogously when considering the Laplace transform.

The Factional Diffusion Equation

We are now going to see how continuous time random walks (CTRW) leads to a fractional diffusion equation.

A CTRW is based on the fact that the length jump and the waiting time between two successive jumps are given by the probability density function – pdf – $\psi(x,t)$. From this we can define the jump length pdf $\lambda(x)$ from the marginal probability

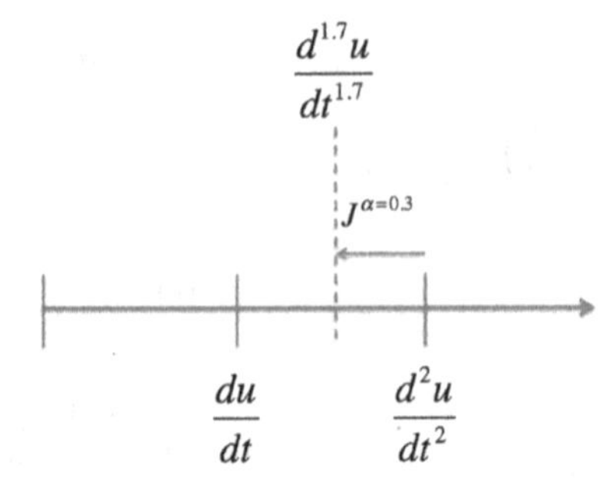

Fig. 3.9 Schematic cartoon of the derivative of order 1.7 (color figure online)

$$\lambda(x) = \int_0^{\infty} \psi(x, t)dt, \tag{3.55}$$

and the waiting time pdf by

$$\omega(t) = \int_{-\infty}^{\infty} \psi(x, t)dx. \tag{3.56}$$

Thus, CTRW can be characterized by the characteristic waiting time T

$$T = \int_0^{\infty} t\omega(t)dt, \tag{3.57}$$

and the jump length variance Σ^2

$$\Sigma^2 = \int_{-\infty}^{\infty} x^2\lambda(x)dx, \tag{3.58}$$

which can be either finite or diverging.

Now, we write the master equation

$$\eta(x, t) = \int_{-\infty}^{\infty} ds \int_0^t \eta(s, \tau)\psi(x - s, t - \tau)d\tau + g(x)\delta(t), \tag{3.59}$$

where the last term accounts for the initial condition, with a distribution $g(x)$ at the initial time $t = 0$. When all the probability concentrates at the origin of coordinates, $g(x) = \delta(x)$, with $\delta(\bullet)$ the Dirac mass. The previous master equation relates the pdf

$\eta(x, t)$ of just arriving at location x at time t to the one of having arrived at location s at time τ.

Thus, the probability of being at position x at time t, $W(x, t)$, is given by

$$W(x,t) = \int_0^t \eta(x,\tau)\Upsilon(t-\tau)d\tau, \tag{3.60}$$

where $\Upsilon(t-\tau)$ represents the probability that no jump takes place in the time interval $[t-\tau, t]$, and can be calculated from

$$\Upsilon(t) = 1 - \int_0^t \omega(\tau)d\tau, \tag{3.61}$$

whose Laplace transform reads as

$$\Upsilon(u) = \frac{1-\omega(u)}{u}. \tag{3.62}$$

Equation (3.60) can also be written as

$$W(x,t) = \int_0^t \eta(x,t-\tau)\Upsilon(\tau)d\tau. \tag{3.63}$$

By introducing Eq. (3.59) into Eq. (3.63), it results that

$$\begin{aligned} W(x,t) &= \int_0^t \left(\int_{-\infty}^{\infty} ds \int_0^{t-\tau} \eta(s,\theta)\psi(x-s,t-\tau-\theta)d\theta + g(x)\delta(\tau) \right) \Upsilon(\tau)d\tau \\ &= \int_0^t \left(\int_{-\infty}^{\infty} ds \int_0^{t-\tau} \eta(s,\theta)\psi(x-s,t-\tau-\theta)d\theta \right) \Upsilon(\tau)d\tau + g(x)\Upsilon(t) \\ &= \int_{-\infty}^{\infty} ds \int_0^t d\tau \left(\int_0^{t-\tau} \eta(s,\theta)\psi(x-s,t-\tau-\theta)d\theta \right) \Upsilon(\tau) + g(x)\Upsilon(t). \end{aligned} \tag{3.64}$$

On the other hand,

$$
\begin{aligned}
&\int_{-\infty}^{\infty} ds \int_{0}^{t} d\tau W(s,\tau)\psi(x-s,t-\tau) \\
&= \int_{-\infty}^{\infty} ds \int_{0}^{t} d\tau \left(\int_0^\tau \eta(\tau-\theta)\Upsilon(\theta)d\theta\right)\psi(x-s,t-\tau) \\
&= \int_{-\infty}^{\infty} ds \int_{0}^{t} d\theta \left(\int_\theta^t \eta(\tau-\theta)\psi(x-s,t-\tau)d\tau\right)\Upsilon(\theta).
\end{aligned}
\tag{3.65}
$$

By defining $\tau - \theta = u$, the previous integral (3.65) reads as

$$
\begin{aligned}
&\int_{-\infty}^{\infty} ds \int_{0}^{t} d\theta \left(\int_\theta^t \eta(\tau-\theta)\psi(x-s,t-\tau)d\tau\right)\Upsilon(\theta) \\
&= \int_{-\infty}^{\infty} ds \int_{0}^{t} d\theta \left(\int_0^{t-\theta} \eta(u)\psi(x-s,t-\theta-u)du\right)\Upsilon(\theta),
\end{aligned}
\tag{3.66}
$$

which coincides with the integral part of expression (3.64). Thus, injecting Eq. (3.66) into Eq. (3.64), we obtain the master equation

$$
W(x,t) = \int_{-\infty}^{\infty} ds \int_{0}^{t} d\tau W(s,\tau)\psi(x-s,t-\tau) + g(x)\Upsilon(t), \tag{3.67}
$$

which by applying both Laplace and Fourier transforms, reduces to

$$
W(k,u) = W(k,u)\psi(k,u) + \Upsilon(u)W_0(k), \tag{3.68}
$$

with $W_0(x)$ the Fourier transform of the initial condition, and with $W_0(x) = 1$ when $g(x) = \delta(x)$. From Eq. (3.68), it results that

$$
W(k,u) = \frac{\Upsilon(u)}{1-\psi(k,u)} W_0(k), \tag{3.69}
$$

which taking into account Eq. (3.62) reads as

$$
W(k,u) = \frac{W_0(k)}{1-\psi(k,u)} \frac{1-\omega(u)}{u}, \tag{3.70}
$$

which represents the desired solution in the Fourier/Laplace spaces.

3.2.2.3 Long Rests Versus Long Jumps: Subdiffusion and Levy Flights

We consider now different cases of CTRW with decoupled jump pdf, i.e., $\psi(x,t) = \omega(t)\lambda(x)$. If both the characteristic waiting time T and the jump length variance Σ^2 are finite, the long-time limit corresponds to a Brownian motion.

If we consider a Poissonian waiting time pdf

$$\omega(t) = \frac{1}{\tau}e^{-\frac{t}{\tau}}, \tag{3.71}$$

with $T = \tau$, and a Gaussian jump length pdf $\lambda(x)$

$$\lambda(x) = \frac{1}{\sqrt{4\pi\sigma^2}}e^{-\frac{x^2}{4\sigma^2}}, \tag{3.72}$$

leading to $\Sigma^2 = 2\sigma^2$, then the Laplace and Fourier transforms of $\omega(t)$ and $\lambda(x)$, respectively, read as

$$\begin{cases} \omega(u) = 1 - u\tau + \Theta(\tau^2) \\ \lambda(k) = 1 - \sigma^2 k^2 + \Theta(k^4) \end{cases}. \tag{3.73}$$

Thus, the jump pdf can be approximated at the lowest orders from

$$\psi(k,u) \approx 1 - u\tau - \sigma^2 k^2. \tag{3.74}$$

In fact, as proven in [11], any pair of pdfs leading to finite T and Σ^2 leads to the same results at lower orders, and thus in the long-time limit.

Introducing the expansions (3.73) and (3.74) into Eq. (3.70) results in

$$W(k,u) = \frac{W_0(k)}{u + \mathcal{K}k^2}, \tag{3.75}$$

with $\mathcal{K} = \frac{\sigma^2}{\tau}$.

Now, from the Fourier and Laplace transforms properties, and in particular

$$\mathcal{F}\left\{\frac{\partial^2 W(x,t)}{\partial x^2}\right\} = -k^2 W(k,t) \tag{3.76}$$

and

$$\mathcal{L}\left\{\frac{\partial W(x,t)}{\partial t}\right\} = uW(x,u) - W_0(x), \tag{3.77}$$

reorganizing Eq. (3.75)

$$uW(k,u) - W_0(k) + \mathcal{K}k^2 W(k,u) = 0, \tag{3.78}$$

we can identify the diffusion equation

$$\frac{\partial W}{\partial t} = \mathscr{K} \frac{\partial^2 W}{\partial x^2}. \tag{3.79}$$

Long Rest and Subdiffusion

We consider a long-tailed waiting time pdf whose characteristic time T diverges,

$$\omega(t) = A \left(\frac{\tau}{t} \right)^{1+\alpha}, \tag{3.80}$$

with $0 < \alpha < 1$, whose Laplace transform writes

$$\omega(u) \approx 1 - (u\tau)^{\alpha}, \tag{3.81}$$

which injected into Eq. (3.70), leads to

$$W(k, u) = \frac{\frac{W_0(k)}{u}}{1 + \mathscr{K}_{\alpha} u^{-\alpha} k^2}, \tag{3.82}$$

with $\mathscr{K}_{\alpha} = \frac{\sigma^2}{\tau^{\alpha}}$.

Taking into account the Laplace transform applied to fractional derivatives, in particular,

$$\mathscr{L}\left\{ D^{-p} W(x, t) \right\} = u^{-p} W(x, u), \tag{3.83}$$

for $p \geq 0$, it results that

$$W(x, t) - W_0(x) = D^{-\alpha} \left(\mathscr{K}_{\alpha} \frac{\partial^2 W(x, t)}{\partial x^2} \right). \tag{3.84}$$

By taking the time derivative to Eq. (3.84), it results that

$$\frac{\partial W(x, t)}{\partial t} = D^{1-\alpha} \left(\mathscr{K}_{\alpha} \frac{\partial^2 W(x, t)}{\partial x^2} \right), \tag{3.85}$$

however, a more valuable expression consists in removing the time derivatives in the right-hand side of Eq. (3.84). For that purpose, it suffices to apply a fractional derivative D^{α} to both members. It is important to note that the Riemann–Liouville fractional derivative of a constant does not vanish, in fact, it is given by $D^{\alpha} 1 = \frac{1}{\Gamma(1-\alpha)} t^{-\alpha}$. Thus, from Eq. (3.84), it results that

$$D^{\alpha} W(x, t) - \frac{t^{-\alpha}}{\Gamma(1-\alpha)} W_0(x) = \mathscr{K}_{\alpha} \frac{\partial^2 W(x, t)}{\partial x^2}. \tag{3.86}$$

The mean squared displacement can be calculated from Eq. (3.75), using the relation

$$\langle x^2 \rangle = \lim_{k \to 0} \left\{ -\frac{\partial^2 W(k,u)}{\partial k^2} \right\} \tag{3.87}$$

and performing the subsequent Laplace inversion, which results in

$$\langle x^2 \rangle = \frac{2\mathcal{K}_\alpha}{\Gamma(1+\alpha)} t^\alpha, \tag{3.88}$$

leading to subdiffusion for $\alpha < 1$ and to standard diffusion as soon as $\alpha = 1$.

Long Jumps and Levy Flights

Finally, instead of considering a diverging waiting time, we consider a diverging jump length variance Σ^2. For the waiting time, we consider the Poissonian pdf, and for the jump length distribution, we consider the Lévy distribution that, in the Fourier space, is expressed from

$$\lambda(k) = e^{-\sigma^\mu |k|^\mu} \approx 1 - \sigma^\mu |k|^\mu, \tag{3.89}$$

for $1 < \mu < 2$, which corresponds to the asymptotic behavior $\lambda(x) \approx A_\mu \sigma^{-\mu} |k|^{-1-\mu}$. In the present case, and as previously indicated, the mean squared displacement diverges.

By injecting expression (3.89) into Eq. (3.70), it results that

$$W(k,u) = \frac{1}{u + \mathcal{K}_\mu |k|^\mu}, \tag{3.90}$$

with $\mathcal{K}_\mu = \frac{\sigma^\mu}{\tau}$. The Laplace and Fourier inversion lead to the fractional equation

$$\frac{\partial W(x,t)}{\partial t} = \mathcal{K}_\mu D^\mu_{-\infty} W(x,t), \tag{3.91}$$

where $D^\mu_{-\infty}$ denotes the fractional derivative previously defined, but with the lower integration limit now being taken at $-\infty$.

Thus, subdiffusion is related to the fractional time derivative, whereas superdiffusion places the fractional derivative in the space derivatives.

References

1. V.B. Shenoy, R. Millera, E.B. Tadmor, D. Rodney, R. Phillips, M. Ortiz, An adaptive finite element approach to atomic-scale mechanics - the quasicontinuum method. J. Mech. Phys. Solids **36**, 500–531 (1999)
2. H. Ben Dhia, Multiscale mechanical problems: the Arlequin method. C. R. Acad. Sci. **326**, 899–904 (1998)

3. G.J. Wagner, W.K. Liu, Coupling of atomistic and continuum simulations using a bridging scale decomposition. J. Comput. Phys. **190**, 249–274 (2003)
4. S.P. Xiao, T. Belytschko, A bridging domain method for coupling continua with molecular dynamics. Comput. Methods Appl. Mech. Eng. **193**, 1645–1669 (2004)
5. P. Español, P. Warren, Statistical mechanics of dissipative particle dynamics. Europhys. Lett. **30**(4), 191–196 (1995)
6. G. Gompper, T. Ihle, D.M. Kroll, R.G. Winkler, Multi-particle collision dynamics: a particle-based mesoscale simulation approach to the hydrodynamics of complex fluids, in *Advanced Computer Simulation Approaches for Soft Matter Sciences III*, ed. by C. Holm, K. Kremer (Springer, Berlin, 2009)
7. J.H. Weiner, *Statistical Mechanics of Elasticity* (Dover, Mineola, 2002)
8. A. Jaishankar, G. McKinley, Power-law rheology in the bulk and at the interface: quasi-properties and fractional constitutive equations. Proc. R. Soc. A **469**, 20120284 (2012)
9. I. Podlubny, *Fractional Differential Equations* (Academic Press, San Diego, 1999)
10. A. Kilbas, H.M. Srivastava, J.J. Trujillo, *Theory and Applications of Fractional Differential Equations* (Elsevier, Amsterdam, 2006)
11. J. Klafter, A. Blumen, M.F. Shlesinger, Stochastic pathway to anomalous diffusion. Phys. Rev. A **35**(7), 3081–3085 (1987)

Chapter 4
Kinetic Theory Models

Abstract Discrete techniques (MD or BD), despite their conceptual simplicity, are very often too expensive from the computational point of view. Kinetic theory approaches seem, in many cases, a suitable compromise between the accuracy of finer descriptions and the computational efficiency of macroscopic descriptions. In this chapter, we revisit some kinetic theory models. Even if there is a common rationale for deriving the different models, in order to emphasize their physical contents, we will follow a diversity of alternative routes to derive them.

Keywords Kinetic theory · Lattice-Boltzmann · Hydrodynamic equations
Fokker-Planck equation · Chemical Master Equation

4.1 Motivation

As indicated above, the atomistic description involves overly rich kinematics. The atoms vibrate (due to thermal effects) and small particles exhibit an erratic path due to the continuous impacts that they endure. Of course, appropriate averages of quantities of physical interest are usually governed by classical deterministic equations.

In the book written by Erwin Schrödinger entitled *What is the life? The physical aspects of the living cells* [1], the author states that the sensorial structures of living beings need a multi-atomic structure to be protected from any random monoatomic event. A living being needs exact physical laws, and for this reason, our sensorial devices must filter the high frequency information (characteristic of atomic behaviors), retaining only averages that evolve with a certain regularity. To perform this filtering process, one needs multiatomic sensorial organs, as found in the vast majority of living beings! In this way, knowledge becomes simple and fast.

Kinetic theory loses individuality in favor of coarser population descriptions. The kinetic theory framework will be described in this chapter by focusing on different physics.

© The Author(s) 2018

F. Chinesta and E. Abisset-Chavanne, *A Journey Around the Different Scales Involved in the Description of Matter and Complex Systems*, SpringerBriefs in Applied Sciences and Technology, https://doi.org/10.1007/978-3-319-70001-4_4

4.2 Kinetic Theory Description of Simple Liquids and Gases

In kinetic theory approaches, we are no longer interested in individual behaviors, but rather in averages. Let N be a population of particles having an electrical charge. The system could be described by giving the position and velocity of each particle in the population at any time, as is the case in particle-based approaches. However, this description is too expensive when we are only interested in averages and not in local effects, such as defaults propagation. In this way, we could introduce a distribution function giving the fraction of particles that, at time t, are located in a neighborhood of point $\mathbf{x}$ ($\mathbf{x} \in \mathbb{R}^3$) and have a velocity close to $\mathbf{v}$ ($\mathbf{v} \in \mathbb{R}^3$).

Thus, $\Psi(\mathbf{x}, \mathbf{v}, t) d\mathbf{x} d\mathbf{v}$ represents the fraction of those particles located inside the hexahedron $[x - \frac{dx}{2}, x - \frac{dx}{2}] \times [y - \frac{dy}{2}, y - \frac{dy}{2}] \times [z - \frac{dz}{2}, z - \frac{dz}{2}]$, whose velocity is also inside the hexahedron $[u - \frac{du}{2}, u - \frac{du}{2}] \times [v - \frac{dv}{2}, v - \frac{dv}{2}] \times [\omega - \frac{d\omega}{2}, \omega - \frac{d\omega}{2}]$, where the components of the position and velocity vectors are given by $\mathbf{x} = (x, y, z)$ and $\mathbf{v} = (u, v, \omega)$, respectively.

Thus, all the information required to describe the system has been compressed into a single scalar function, the probability density function – pdf – $\Psi(\mathbf{x}, \mathbf{v}, t)$, and the only price to be paid concerns its multi-dimensional character.

Despite the high compactness of such a kinetic theory description, it is not useful without an equation governing the evolution of the joint distribution function $\Psi(\mathbf{x}, \mathbf{v}, t)$ from the prescribed initial and boundary conditions. The equation governing the evolution of the distribution function results from the particle conservation balance in the physical and conformational spaces that reads as

$$\frac{\partial \Psi}{\partial t} + \nabla_x \cdot (\mathbf{q}_x) + \nabla_v \cdot (\mathbf{q}_v) = S, \tag{4.1}$$

where $\nabla_x \cdot (\bullet)$ and $\nabla_v \cdot (\bullet)$ denote the divergence linear differential operator in the physical (space) and conformational (velocity) coordinates, respectively, and the source S is called the collision term. The convective fluxes in both spaces, those of the physical $\mathbf{q}_x$ and the conformational $\mathbf{q}_v$, are given by

$$\begin{cases} \mathbf{q}_x = \dot{\mathbf{x}}\ \Psi = \mathbf{v}\Psi \\ \mathbf{q}_v = \dot{\mathbf{v}}\ \Psi = \mathbf{a}\Psi \end{cases}, \tag{4.2}$$

with $\mathbf{a}$ the acceleration field. With this notation, the balance equation reads as

$$\frac{\partial \Psi}{\partial t} + \nabla_x \cdot (\mathbf{v}\Psi) + \nabla_v \cdot (\mathbf{a}\Psi) = S. \tag{4.3}$$

Obviously, before solving Eq. (4.3), one needs to specify both the collision term and the acceleration field, as well as the initial and boundary conditions. The velocity field $\mathbf{v}$ that appears in the second term does not require any specific treatment, because in the present approach, the velocity components are, in fact, coordinates, like space and time. The collision term depends on the physics considered. In turn, the acceleration field can be derived from the fact that the population that we are considering consists of a set of particles with electrical charge that are interacting through the Coulomb potential. In the kinetic theory description, localized charges are substituted with the charge distribution

$$Q(\mathbf{x}, t) = Z \int_{\mathbb{R}^3} \Psi(\mathbf{x}, \mathbf{v}, t) d\mathbf{v}, \tag{4.4}$$

where Z is the total charge in the system, as the following expression proves:

$$\int_{\mathbb{R}^3} Q(\mathbf{x}, t) d\mathbf{x} = Z \int_{\mathbb{R}^3} \left(\int_{\mathbb{R}^3} \Psi(\mathbf{x}, \mathbf{v}, t) d\mathbf{v} \right) d\mathbf{x} = Z, \quad \forall t. \tag{4.5}$$

Now, an electrostatic potential $V(\mathbf{x}, t)$ can be associated with this charge distribution $Q(\mathbf{x}, t)$, which verifies the elliptic partial differential equation

$$\nabla_x^2 V(\mathbf{x}, t) = Q(\mathbf{x}, t); \quad \forall t, \tag{4.6}$$

where ∇_x denotes the gradient linear differential operator with respect to the space coordinates $\mathbf{x}$. From the electrostatic potential that results from the solution of Eq. (4.6), we can compute the electrical field $\mathbf{E}(\mathbf{x}, t)$

$$\mathbf{E}(\mathbf{x}, t) = -\nabla_x V(\mathbf{x}, t), \tag{4.7}$$

from which we can easily derive the expression of the force $\mathbf{F}(\mathbf{x}, t)$ acting on a volume $d\Omega$ located at point $\mathbf{x}$ (at time t)

$$\mathbf{F}(\mathbf{x}, t) = \mathbf{E}(\mathbf{x}, t) Q(\mathbf{x}, t) d\Omega, \tag{4.8}$$

which gives the acceleration

$$a(\mathbf{x}, t) = \frac{\mathbf{F}(\mathbf{x}, t)}{\rho(\mathbf{x}, t) d\Omega} = -\nabla_x V(\mathbf{x}, t) \frac{Q(\mathbf{x}, t)}{\rho(\mathbf{x}, t)}, \tag{4.9}$$

where $\rho(\mathbf{x}, t)$ is the density,

$$\rho(\mathbf{x}, t) = M \int_{\mathbb{R}^3} \Psi(\mathbf{x}, \mathbf{v}, t) d\mathbf{v}, \tag{4.10}$$

with M the total mass in the system.

Thus, the model can be summarized by the following equations:

$$\begin{cases} \frac{\partial \Psi}{\partial t} + \nabla_x \cdot (\mathbf{v}\Psi) + \nabla_v \cdot (\mathbf{a}\Psi) = S \\ \mathbf{a}(\mathbf{x}, t) = -\nabla_x V(\mathbf{x}, t) \frac{Q(\mathbf{x},t)}{\rho(\mathbf{x},t)} \\ Q(\mathbf{x}, t) = Z \int_{\mathbb{R}^3} \Psi(\mathbf{x}, \mathbf{v}, t) d\mathbf{v} \\ \rho(\mathbf{x}, t) = M \int_{\mathbb{R}^3} \Psi(\mathbf{x}, \mathbf{v}, t) d\mathbf{v} \\ \nabla_x^2 V(\mathbf{x}, t) = Q(\mathbf{x}, t) \end{cases} . \tag{4.11}$$

This model has been widely applied for modeling quantum gases (plasma) and it is known as the Vlasov–Poisson–Boltzmann model.

If the particles are not charged, the acceleration vanishes, i.e., $\mathbf{a}(\mathbf{x}, t) = \mathbf{0}$, except during the instantaneous collisions, then the steady solution results in the Maxwell–Boltzmann distribution that is considered to be the equilibrium distribution Ψ_{eq}.

In fact, the equilibrium distribution allows us to define, within the so-called BFK (Bhatnagar–Gross–Krook) model, the collision term as

$$S = -\nu_{ref}(\Psi - \Psi_{eq}), \tag{4.12}$$

where ν_{eq} is the inverse of a characteristic time. Now, in absence of electrical charge, the equation governing the evolution of the distribution could be written in the form:

$$\frac{\partial \Psi}{\partial t} + \nabla_x \cdot (\mathbf{v}\Psi) = -\nu_{eq}(\Psi - \Psi_{eq}). \tag{4.13}$$

From this analysis, we can conclude that the use of extremely large populations of particles characteristic of discrete simulation techniques (MD or BD) can be replaced with the solution of a continuous model. However, the solution of the resulting partial differential equation is a tricky issue because of its hyperbolic character and its multidimensionality. In [2], authors considered the use of separate representations for alleviating the difficulties to which we have just referred.

Lattice-Boltzmann approaches can be viewed as an attempt to circumvent both difficulties, and they have attracted the interest of many researchers over recent decades. A brief summary is given in Sect. 4.2.2, and for additional details, the interested reader can refer to [3] and the references therein.

4.2.1 Hydrodynamic Equations

When considering the Boltzmann equation

$$\frac{\partial \Psi}{\partial t} + \nabla_x \cdot (\mathbf{v}\Psi) = S(\mathbf{x}, \mathbf{v}, t), \tag{4.14}$$

its solution $\Psi(\mathbf{x}, \mathbf{v}, t)$ allows us to define the macroscopic quantities $\rho(\mathbf{x}, t)$ and $\mathbf{u}(\mathbf{x}, t)$, the density and the macroscopic velocity, respectively, from

$$\rho(\mathbf{x}, t) = \int_{\mathbb{R}^3} \Psi(\mathbf{x}, \mathbf{v}, t) d\mathbf{v}, \tag{4.15}$$

and

$$\rho(\mathbf{x}, t)\mathbf{u}(\mathbf{x}, t) = \int_{\mathbb{R}^3} \mathbf{v}\Psi(\mathbf{x}, \mathbf{v}, t) d\mathbf{v}. \tag{4.16}$$

The pressure tensor $\mathbf{P}$, which contains the scalar pressure and the viscous stress, results from

$$\mathbf{P}(\mathbf{x}, t) = \int_{\mathbb{R}^3} (\mathbf{v} - \mathbf{u}(\mathbf{x}, t)) \otimes (\mathbf{v} - \mathbf{u}(\mathbf{x}, t)) \, \Psi(\mathbf{x}, \mathbf{v}, t) d\mathbf{v}, \tag{4.17}$$

which leads to

$$\mathbf{P}(\mathbf{x}, t) = -\rho(\mathbf{x}, t) \, \mathbf{u}(\mathbf{x}, t) \otimes \mathbf{u}(\mathbf{x}, t) + \int_{\mathbb{R}^3} \mathbf{v} \otimes \mathbf{v} \, \Psi(\mathbf{x}, \mathbf{v}, t) d\mathbf{v}. \tag{4.18}$$

The mass and momentum conservation imply that

$$\begin{cases} \int_{\mathbb{R}^3} S(\mathbf{x}, \mathbf{v}, t) d\mathbf{v} = 0 \\ \int_{\mathbb{R}^3} \mathbf{v} S(\mathbf{x}, \mathbf{v}, t) d\mathbf{v} = 0 \end{cases}. \tag{4.19}$$

Now, in order to obtain the first hydrodynamic equation, it suffices to integrate the Boltzmann equation as follows:

$$\int_{\mathbb{R}^3} \left(\frac{\partial \Psi}{\partial t} + \nabla_x \cdot (\mathbf{v}\Psi) - S \right) d\mathbf{v} = 0, \tag{4.20}$$

implaying that

$$\frac{\partial \rho}{\partial t} + \nabla_x \cdot (\rho \mathbf{u}) = 0, \tag{4.21}$$

which corresponds to the standard continuity equation. Now, by multiplying the Boltzmann equation by $\mathbf{v}$ and then integrating it into the velocity space, it results that

$$\int_{\mathbb{R}^3} \mathbf{v} \left(\frac{\partial \Psi}{\partial t} + \nabla_x \cdot (\mathbf{v}\Psi) - S \right) d\mathbf{v} = \mathbf{0}, \tag{4.22}$$

which operating and making use of the previous definitions, becomes

$$\frac{\partial (\rho \mathbf{u})}{\partial t} + \nabla_x \cdot (\rho \mathbf{u} \otimes \mathbf{u}) + \nabla_x \cdot \mathbf{P} = \mathbf{0}. \tag{4.23}$$

In order to make explicit the macroscopic expression of the pressure tensor $\mathbf{P}$, the simplest route consists in using the Chapman–Enskog expansion that expresses the distribution function $\Psi(\mathbf{x}, \mathbf{v}, t)$ from the asymptotic expansion

$$\Psi = \sum_{i=0}^{\infty} \varepsilon^i \Psi^{(i)}. \tag{4.24}$$

Then, this expansion is injected into the Boltzmann equation using the BFK model of the collision term according to

$$\frac{\partial(\Psi^{(0)} + \varepsilon\Psi^{(1)} + \cdots)}{\partial t} + \mathbf{v} \cdot \nabla_x \left(\Psi^{(0)} + \varepsilon\Psi^{(1)} + \cdots\right)$$
$$= \frac{\left(\Psi^{(0)} + \varepsilon\Psi^{(1)} + \cdots\right) - \Psi^{eq}}{\tau}, \tag{4.25}$$

with $\varepsilon\Theta\tau$. By identifying the terms at the different orders, we obtain:

- Order ε^{-1}:

$$\Psi^{(0)} = \Psi^{eq}. \tag{4.26}$$

- Order ε^0:

$$\frac{\partial\Psi^{(0)}}{\partial t} + \mathbf{v} \cdot \nabla_x\Psi^{(0)} = \Psi^{(1)}. \tag{4.27}$$

- Order ε^1:

$$\frac{\partial\Psi^{(1)}}{\partial t} + \mathbf{v} \cdot \nabla_x\Psi^{(1)} = \Psi^{(2)}, \tag{4.28}$$

  and so on for the next orders.

The first order approximation of the pressure tensor $\mathbf{P}^{(0)}$ is obtained by considering the zero order distribution function $\Psi^{(0)}$ that, according to the Chapman–Enskog expansion just discussed, corresponds to the equilibrium Maxwell–Boltzmann distribution $\Psi^{eq}(\mathbf{v})$, and consequently

$$\mathbf{P}^{(0)}(\mathbf{x}, t) = \int_{\mathbb{R}^3} (\mathbf{v} - \mathbf{u}(\mathbf{x}, t)) \otimes (\mathbf{v} - \mathbf{u}(\mathbf{x}, t))\, \Psi^{eq}(\mathbf{v}) d\mathbf{v} = K_b T \rho(\mathbf{x}, t), \tag{4.29}$$

which injected into Eq. (4.23), leads to the Euler equation.

The second order approximation proceeds from

$$\mathbf{P}^{(1)}(\mathbf{x}, t) = \int_{\mathbb{R}^3} (\mathbf{v} - \mathbf{u}(\mathbf{x}, t)) \otimes (\mathbf{v} - \mathbf{u}(\mathbf{x}, t))\, \Psi^{(1)}(\mathbf{x}, \mathbf{v}, t) d\mathbf{v}, \tag{4.30}$$

which according to the Chapman–Enskog expansion just discussed, results in

$$\Psi^{(1)} = \frac{\partial \Psi^{(0)}}{\partial t} + \mathbf{v} \cdot \nabla_x \Psi^{(0)}, \tag{4.31}$$

with $\Psi^{(0)} = \Psi^{eq}$. It finally results that

$$\mathbf{P}^{(1)}(\mathbf{x}, t) = 2K_b T \rho \tau (\nabla_x \mathbf{u} + (\nabla_x \mathbf{u})^T). \tag{4.32}$$

Defining the viscosity $\mu = K_b T \rho \tau$, and considering $\mathbf{P} = \mathbf{P}^{(0)} + \mathbf{P}^{(1)}$ in Eq. (4.23), the Navier–Stokes equation is obtained.

4.2.2 The Lattice Boltzmann Method – LBM

Use of the LBM provides many of the advantages of molecular dynamics, including clear physical pictures, easy implementation, and fully parallel algorithms. The basic premise for using this simplified kinetic-type method for macroscopic fluid flows is that the macroscopic dynamics of a fluid is the result of the collective behavior of many microscopic particles in the system and that the macroscopic dynamics is not sensitive to the underlying details in microscopic physics. The LBM's ancestor is the lattice gas automata, which is constructed as a simplified, fictitious molecular dynamics in which space, time, and particle velocities are all discrete. It consists of a regular lattice with particles residing on the nodes. Starting from an initial state, the configuration of particles at each time step evolves in two sequential sub-steps: (i) streaming, in which each particle moves to the nearest node in the direction of its velocity, and (ii) collision, which occurs when particles arriving at a node interact and change their velocity directions according to specified rules. The Lattice Boltzmann method considers a number of velocity directions depending on the considered cell and a velocity value associated with each of these directions, allowing it to reach the neighbor location in the considered time step.

We consider the Boltzmann equation within the BFK (Bhatnagar–Gross–Krook) modeling of the collision term,

$$\frac{\partial \Psi}{\partial t} + \nabla_x \cdot (\mathbf{v}\Psi) = -\nu_{eq}(\Psi - \Psi_{eq}), \tag{4.33}$$

and consider a regular grid, the so-called lattice, of size Δx. For the sake of simplicity, we consider the two-dimensional model in which the particles are constrained to occupy the nodes of the lattice. This fact constitutes the first simplification. The second assumption restricts a particle to streaming in a finite number of possible directions, 9 when considering the lattice node illustrated in Fig. 4.1, usually referred as D2Q9. These velocities are denoted by vectors $\mathbf{e}_i$, $i = 1, \ldots, 8$. The reduction accomplished is quite impressive: an infinity of points $\mathbf{x}$, each of them having an

Fig. 4.1 D2Q9 lattice node

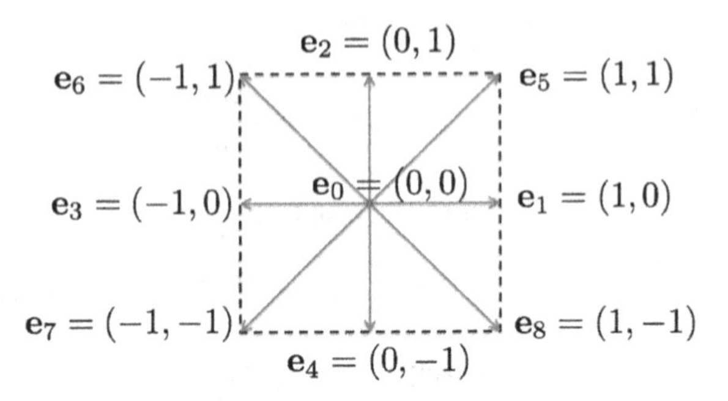

infinity of directions and magnitude of the possible velocities, has been reduced to the nodes in the lattice, with only 8 possible directions and only one velocity magnitude associated with each direction, 1 for directions 1 to 4 and $\sqrt{2}$ for directions 5 to 8.

Now, we denote by $\Psi_i(\mathbf{x}, t)$ the probability of streaming in direction $\mathbf{e}_i$ with its associated velocity at the lattice point $\mathbf{x}$ and time t. The two steps in the Lattice-Boltzmann simulations are the streaming and the collision, which combined read as, $\forall\, i$

$$\Psi_i(\mathbf{x} + c\Delta t\mathbf{e}_i, t + \Delta t) - \Psi_i(\mathbf{x}, t) = -\nu_{eq}\left(\Psi_i(\mathbf{x}, t) - \Psi_i^{eq}(\mathbf{x}, t)\right), \tag{4.34}$$

where the left-hand side represents the streaming, the right-hand side the collision, and c is the so-called lattice speed, $c = \frac{\Delta x}{\Delta t}$.

The macroscopic density ρ and velocity $\mathbf{u}$ read as

$$\rho(\mathbf{x}, t) = \sum_{i=0}^{8} \Psi_i(\mathbf{x}, t), \tag{4.35}$$

and

$$\mathbf{u}(\mathbf{x}, t) = \frac{1}{\rho(\mathbf{x}, t)} \sum_{i=0}^{8} c\Psi_i(\mathbf{x}, t)\mathbf{e}_i, \tag{4.36}$$

respectively.

Now, when considering single phase flows, the D2Q9 lattice and the BGK collision term, the equilibrium distribution Ψ_i^{eq} reads as

$$\Psi_i^{eq} = \rho\omega_i + \rho s_i(\mathbf{u}(\mathbf{x}, t)), \tag{4.37}$$

where

$$s_i(\mathbf{u}) = \omega_i\left(3\frac{\mathbf{e}_i \cdot \mathbf{u}}{c} + \frac{9}{2}\frac{(\mathbf{e}_i \cdot \mathbf{u})^2}{c^2} - \frac{3}{2}\frac{\mathbf{u} \cdot \mathbf{u}}{c^2}\right), \tag{4.38}$$

with the weights given by

$$\omega_i = \begin{cases} \frac{4}{9} & i = 0 \\ \frac{1}{9} & i = 1, \ldots, 4 \\ \frac{1}{36} & i = 5, \ldots, 8 \end{cases} . \tag{4.39}$$

To obtain these expressions, it suffices to define an equilibrium distribution that fulfills the macroscopic definitions $\sum_i \Psi_i^{eq} = \rho$, $\sum_i \Psi_i^{eq}(c\mathbf{e}_i - \mathbf{u}) = \mathbf{0}$, etc. and from them, to identify the different terms Ψ_i^{eq}.

The simulation algorithm can be summarized as follows:

1. Initialize Ψ_i, and from it compute ρ, $\mathbf{u}$, and then Ψ_i^{eq};
2. *Streaming step*: From Ψ_i, compute $\Psi_i^*(\mathbf{x} + c\Delta t\mathbf{e}_i, t + \Delta t) = \Psi(\mathbf{x}, t)$;
3. From Ψ_i^*, compute ρ and $\mathbf{u}$, and from both, Ψ_i^{eq};
4. *Collision step*: Update the distribution Ψ_i from $\Psi_i = \Psi_i^* - \nu(\Psi_i^* - \Psi_i^{eq})$.

The main issues related to the use of this procedure (very simple at first glance) are the enforcement of Dirichlet boundary conditions and the calculation of the expression of the equilibrium distribution for complex or multi-phase flows. The interested reader can refer to the vast existing bibliography on these topics.

4.3 Kinetic Theory Description of Some Complex Fluids

Over recent decades, an increasing number of functional and structural parts, so far made with metals, has been progressively reengineered by replacing metallic materials with polymers, reinforced polymers and composites. The motivation for this substitution may be the weight reduction, the simpler, cheaper or faster forming process, or the ability to exploit additional functionalities. The fillers usually employed cover a broad range, involving many scales: (i) the nanometer scale (e.g., carbon nanotubes, graphene, fullerene, nanodiamonds); (ii) the micrometer to the millimeter scale (particles and short fibers); (iii) the centimeter scale of fibers used in SMC and BMC composite processes; and finally, (iv) the macroscopic scale in which fibrous reinforcements are made of continuous fibers arranged in bundles. When load-bearing capacities are especially sought after, continuous fiber reinforced polymers are selected. In that case, the impregnation of the reinforcement with a low viscosity polymer involves the flow of a Newtonian or non-Newtonian fluid through the complex multi-scale microstructure related to the fiber and tow arrangement. Reinforced polymers are selected instead of high performance polymers of equivalent properties, since the latter are generally more expensive. When looking for functional properties, the use of nano-charges opens up a wide spectrum of possibilities but also raises new challenges, such as dispersion of charges into the polymer matrix and occurrence of aggregation and disaggregation mechanisms. Suspensions of practical interest involve many scales and many concentration regimes, the latter ranging from diluted to highly concentrated.

In what follows, we address many of the scenarios to which we have just referred. We start by addressing the modeling of suspensions involving rigid rods in which the Brownian effects are neglected. For this purpose, we consider a suspending medium consisting of a Newtonian fluid of viscosity η in which there are suspended rigid and non-Brownian rods. We assume, as a first approximation, that their presence and orientation do not affect the flow kinematics that is defined by the velocity field $\mathbf{v}(\mathbf{x}, t)$, which allows for defining the velocity gradient, rate of strain and vorticity tensors, $\nabla \mathbf{v}$, $\mathbf{D}$ and $\boldsymbol{\Omega}$, respectively, with $\mathbf{D}$ and $\boldsymbol{\Omega}$ the symmetric and skew-symmetric components of the velocity gradient, respectively, i.e.,

$$\mathbf{D} = \frac{1}{2}\left(\nabla \mathbf{v} + (\nabla \mathbf{v})^T\right) \tag{4.40}$$

and

$$\boldsymbol{\Omega} = \frac{1}{2}\left(\nabla \mathbf{v} - (\nabla \mathbf{v})^T\right). \tag{4.41}$$

In what follows, we consider the following tensor products:

- if $\mathbf{a}$ and $\mathbf{b}$ are first order tensors, then the single contraction "$\cdot$" reads as $(\mathbf{a} \cdot \mathbf{b}) = a_j\, b_j$ (Einstein summation convention);
- if $\mathbf{a}$ and $\mathbf{b}$ are first order tensors, then the dyadic product "$\otimes$" reads as $(\mathbf{a} \otimes \mathbf{b})_{jk} = a_j\, b_k$;
- if $\mathbf{a}$ and $\mathbf{b}$ are first order tensors, then the cross product "$\times$" reads as $(\mathbf{a} \times \mathbf{b})_j = \varepsilon_{jmn}\, a_m\, b_n$ (Einstein summation convention) with ε_{jmn} the components of the Levi-Civita tensor $\boldsymbol{\varepsilon}$ (also known as a permutation tensor);
- if $\mathbf{a}$ and $\mathbf{b}$ are, respectively, second and first order tensors, then the single contraction "$\cdot$" reads as $(\mathbf{a} \cdot \mathbf{b})_j = a_{jm}\, b_m$ (Einstein summation convention);
- if $\mathbf{a}$ and $\mathbf{b}$ are second order tensors, then the single contraction "$\cdot$" reads as $(\mathbf{a} \cdot \mathbf{b})_{jk} = a_{jm}\, b_{mk}$ (Einstein summation convention);
- if $\mathbf{a}$ and $\mathbf{b}$ are second order tensors, then the double contraction ':" reads as $(\mathbf{a} : \mathbf{b}) = a_{jk}\, b_{kj}$ (Einstein summation convention).

4.3.1 Dilute Suspensions of Non-Brownian Rods

To address the modeling of suspensions involving rigid rods while neglecting the Brownian effects, we consider nine different bricks that constitute the multiscale description of such suspensions: three at the microscopic scale, three at the mesoscopic level and three at the macroscopic scale, all them revisited in what follows.

Microscopic Scale: Particle Conformation

The conformation of each rod of length $2L$ can be described from its orientation, the last expressed from the unit vector $\mathbf{p}$ located at the rod center of gravity $\mathbf{G}$ and aligned along the rod axis. Despite the fact that we are using the concept of rod's center of gravity, the rod mass can be neglected, and with it all the inertia effects.

Microscopic Scale: Particle Conformation Evolution

The equation governing the time evolution of the particle conformation $\dot{\mathbf{p}}$ can be derived by considering the system illustrated in Fig. 4.2 consisting of a rod and two beads located at both ends of the rod where we assume that hydrodynamic forces act. We assume that the forces $\mathbf{F}$ that act on each bead scale with the difference of velocities between the fluid and the bead, the first one given by $\mathbf{v}_0 + \nabla\mathbf{v} \cdot \mathbf{p}L$ and the second one by $\mathbf{v}_G + \dot{\mathbf{p}}L$. Thus, force $\mathbf{F}(\mathbf{p}L)$ reads as

$$\mathbf{F}(\mathbf{p}L) = \xi(\mathbf{v}_0 + \nabla\mathbf{v} \cdot \mathbf{p}L - \mathbf{v}_G - \dot{\mathbf{p}}L), \tag{4.42}$$

where ξ is the friction coefficient, $\mathbf{v}_0$ the fluid velocity at the rod's center of gravity (assumed unperturbed by the rod presence and orientation) and $\mathbf{v}_G$ the velocity of the center of gravity.

Obviously if $\mathbf{F}(\mathbf{p}L)$ acts on the bead $\mathbf{p}L$, then in the opposite bead $-\mathbf{p}L$, the resulting force reads as

$$\mathbf{F}(-\mathbf{p}L) = \xi(\mathbf{v}_0 - \nabla\mathbf{v} \cdot \mathbf{p}L - \mathbf{v}_G + \dot{\mathbf{p}}L). \tag{4.43}$$

Adding Eqs. (4.42) and (4.43) and enforcing the forces balance neglecting the effects of inertia, (the rod's mass is assumed negligible), it results that

$$\mathbf{F}(\mathbf{p}L) + \mathbf{F}(-\mathbf{p}L) = 2\xi(\mathbf{v}_0 - \mathbf{v}_G) = \mathbf{0}, \tag{4.44}$$

which implies that $\mathbf{v}_0 = \mathbf{v}_G$, that is, the rod's center of gravity is moving with the fluid velocity. To simplify the notation, from now on, we consider $\mathbf{F} = \mathbf{F}(\mathbf{p}L)$ and $\mathbf{F}(-\mathbf{p}L) = -\mathbf{F}$.

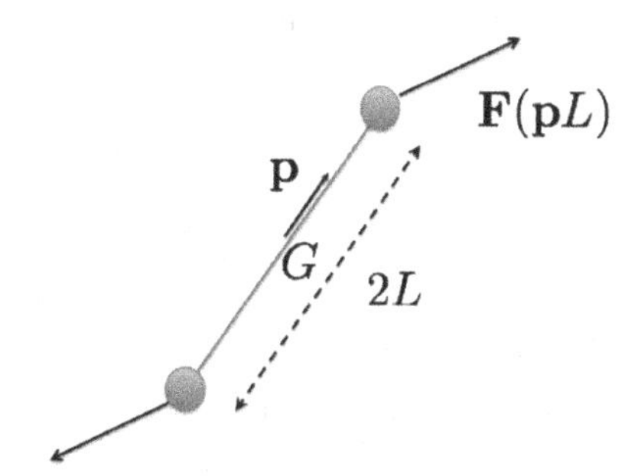

Fig. 4.2 Hydrodynamic forces applied on a rod immersed in a Newtonian fluid

As the resulting torque must also vanish, the only possibility is, that force $\mathbf{F}$ acts along $\mathbf{p}$, that is, $\mathbf{F} = \lambda \mathbf{p}$, with $\lambda \in \mathbb{R}$. Thus, we can write

$$\lambda \mathbf{p} = \xi L(\nabla \mathbf{v} \cdot \mathbf{p} - \dot{\mathbf{p}}). \tag{4.45}$$

Premultiplying Eq. (4.45) by $\mathbf{p}$ and taking into account that $\mathbf{p} \cdot \mathbf{p} = 1$, and consequently $\mathbf{p} \cdot \dot{\mathbf{p}} = 0$, it results that

$$\lambda = \xi L \left(\nabla \mathbf{v} : (\mathbf{p} \otimes \mathbf{p}) \right), \tag{4.46}$$

where $\nabla \mathbf{v} : (\mathbf{p} \otimes \mathbf{p}) \equiv \mathbf{p}^T \cdot \nabla \mathbf{v} \cdot \mathbf{p}$ (both expressions $\nabla \mathbf{v} : (\mathbf{p} \otimes \mathbf{p})$ and $\mathbf{p}^T \cdot \nabla \mathbf{v} \cdot \mathbf{p}$ will be used in what follows in different ways).

Coming back to Eq. (4.45) and using Eq. (4.46), we obtain

$$\xi L \left(\nabla \mathbf{v} : (\mathbf{p} \otimes \mathbf{p}) \right) \mathbf{p} = \xi L(\nabla \mathbf{v} \cdot \mathbf{p} - \dot{\mathbf{p}}), \tag{4.47}$$

from which it ultimately results that the rotary velocity corresponds to the Jeffery equation for infinite aspect ratio ellipsoids (rods) [4]

$$\dot{\mathbf{p}} = \nabla \mathbf{v} \cdot \mathbf{p} - \left(\nabla \mathbf{v} : (\mathbf{p} \otimes \mathbf{p}) \right) \mathbf{p}, \tag{4.48}$$

or its equivalent form

$$\dot{\mathbf{p}} = \nabla \mathbf{v} \cdot \mathbf{p} - \left(\mathbf{p}^T \cdot \nabla \mathbf{v} \cdot \mathbf{p} \right) \mathbf{p}, \tag{4.49}$$

where it can be noticed that the rod's kinematics does not contain size effects.

Microscopic Scale: Particle Contribution to the Stress

The forces acting at the rod's ends $\mathbf{p}L$ and $-\mathbf{p}L$ result, respectively, in $\lambda \mathbf{p}$ and $-\lambda \mathbf{p}$, both directed along the rods direction and auto-equilibrated by construction.

With λ given by Eq. (4.46) it results that

$$\mathbf{F}(\mathbf{p}L) = \xi L(\nabla \mathbf{v} : (\mathbf{p} \otimes \mathbf{p}))\, \mathbf{p}, \tag{4.50}$$

whose contribution to the stress $\boldsymbol{\tau}^p$, by applying Kramer's formula (also known as virial stress), results in

$$\boldsymbol{\tau}^p = \xi L^2 (\nabla \mathbf{v} : (\mathbf{p} \otimes \mathbf{p}))\, \mathbf{p} \otimes \mathbf{p}, \tag{4.51}$$

which can be rewritten as

$$\boldsymbol{\tau}^p = \xi L^2\, \nabla \mathbf{v} : (\mathbf{p} \otimes \mathbf{p} \otimes \mathbf{p} \otimes \mathbf{p}). \tag{4.52}$$

From Eq. (4.52), one could infer that rheology contains size effects because of the presence of factor L^2 in the particle contribution to the stress. To prove that size effects also disappears when addressing rheology, we consider the traction applied on the rod-end $\mathbf{T}$ that, by considering a Newtonian behavior, writes $\mathbf{T} = 2\eta \mathbf{D} \cdot \mathbf{p}$. By writing the objective stress related to the rod's rotation $2\eta(\dot{\mathbf{p}} - \boldsymbol{\Omega} \cdot \mathbf{p})$ and enforcing that the resulting force aligns along the direction $\mathbf{p}$, we again obtain Jeffery's equation. In order to obtain the same expression for the stress, it suffices to consider $\xi \propto \frac{\eta}{L^2}$. Thus, we can conclude that size effects are absent in Eq. (4.52).

Mesoscopic Scale: Population Description

There are two natural descriptions of a population of rods at a mesoscopic level:

- The first consists in specifying each rod orientation by considering the unit vector aligned along its axis, that is, by considering $\mathbf{p}_i, i = 1, \ldots, N$. As will be discussed later, the main drawback related to such an approach lies in the necessity of tracking the evolution of each "computational" rod by solving the corresponding Jeffery equation (4.49), and even if conceptually there is no major difficulty, the computing cost could be excessive in most practical applications.
- The second approach for describing such a population lies in the introduction of the probability distribution function – pdf – $\Psi(\mathbf{x}, t, \mathbf{p})$, given the fraction of rods that a position $\mathbf{x}$ and time t are oriented along direction $\mathbf{p}$.

Despite the fact that both mesoscopic models involve the same physics and richness of description, the main advantage of the second one is the manipulation of a continuous scalar function instead of the discrete description involved in the first approach. The price to be paid when using the description based on the use of the pdf is its inherent multidimensionality, because in that framework, the pdf depends on the standard space and time coordinates, $\mathbf{x}$ and t, respectively, and also on the conformation coordinates that the microstructural description involves, in the present case, $\mathbf{p}$.

Mesoscopic Scale: Description of the Population Evolution

- When the population is described according to the individuals composing it, whose conformation is given by vectors $\mathbf{p}_i$, $i = 1, \ldots, N$, the evolution of each one is given by the Jeffery equation (4.49), which reads as

$$\dot{\mathbf{p}}_i = \nabla \mathbf{v} \cdot \mathbf{p}_i + (\nabla \mathbf{v} : (\mathbf{p}_i \otimes \mathbf{p}_i))\, \mathbf{p}_i, \quad \forall i = 1, \ldots, N. \tag{4.53}$$

- The alternative description consists of using the pdf $\Psi(\mathbf{x}, t, \mathbf{p})$ that verifies the normality condition

$$\int_{\mathcal{S}} \Psi(\mathbf{x}, t, \mathbf{p})\, d\mathbf{p} = 1; \quad \forall \mathbf{x}, \ \ \forall t. \tag{4.54}$$

  In order to use it, one needs to derive the equation governing the evolution of the orientation distribution function Ψ, which is related to the rod's conservation balance, which reads as

$$\frac{\partial \Psi}{\partial t} + \nabla_x \cdot (\dot{\mathbf{x}}\, \Psi) + \nabla_p \cdot (\dot{\mathbf{p}}\, \Psi) = 0, \tag{4.55}$$

where for inertialess rods, with length L smaller than the characteristic flow length, $\dot{\mathbf{x}} = \mathbf{v}(\mathbf{x}, t)$, and the rod's rotary velocity is given by Jeffery's equation (4.49). The resulting balance is known as the orientation Fokker–Planck equation.
The price to pay is the increase of the model's dimensionality, because the orientation distribution is defined in a high-dimensional domain consisting of 6 dimensions in the general 3D case, i.e., $\Psi : (\mathbf{x}, t, \mathbf{p}) \to \mathbb{R}^+$.

Mesoscopic Scale: Contribution of the Particles Population to the Stress

Again, we consider the two alternative descriptions:

- When the population is described in a discrete manner, by specifying the different rods' orientations $\mathbf{p}_i$, the contributions of rods to the suspension stress is calculated by adding their individual effects, given by Eq. (4.52), that is,

$$\boldsymbol{\tau} = \sum_{i=1}^{N} \boldsymbol{\tau}^i = \sum_{i=1}^{N} \xi L^2 \, \nabla \mathbf{v} : (\mathbf{p}_i \otimes \mathbf{p}_i \otimes \mathbf{p}_i \otimes \mathbf{p}_i). \tag{4.56}$$

- When the population is described according to the pdf $\Psi(\mathbf{x}, t, \mathbf{p})$ the sum in Eq. (4.56) is replaced by an integral in the conformation space $\mathscr{S}$:

$$\begin{aligned} \boldsymbol{\tau} &= \int_{\mathscr{S}} \boldsymbol{\tau}^p \, \Psi(\mathbf{p}) \, d\mathbf{p} = 2\eta N_p \int_{\mathscr{S}} \nabla \mathbf{v} : (\mathbf{p} \otimes \mathbf{p} \otimes \mathbf{p} \otimes \mathbf{p}) \, \Psi(\mathbf{p}) \, d\mathbf{p} \\ &= 2\eta N_p \nabla \mathbf{v} : \int_{\mathscr{S}} \mathbf{p} \otimes \mathbf{p} \otimes \mathbf{p} \otimes \mathbf{p} \, \Psi(\mathbf{p}) \, d\mathbf{p}, \end{aligned} \tag{4.57}$$

where the so-called particle number N_p accounts for the particle concentration and where we considered the viscosity instead of the friction coefficient to be consistent with the usual notation.
By considering the expression of the fourth order orientation tensor

$$\mathbf{A} = \int_{\mathscr{S}} \mathbf{p} \otimes \mathbf{p} \otimes \mathbf{p} \otimes \mathbf{p} \, \Psi(\mathbf{p}) \, d\mathbf{p}, \tag{4.58}$$

Eq. (4.57) can be rewritten as

$$\boldsymbol{\tau} = 2\eta N_p (\mathbf{A} : \nabla \mathbf{v}), \tag{4.59}$$

and because of the symmetry of tensor $\mathbf{A}$, the extra-stress $\boldsymbol{\tau}$ can be expressed from

$$\boldsymbol{\tau} = 2\eta N_p (\mathbf{A} : \mathbf{D}). \tag{4.60}$$

Macroscopic Scale: Conformation

As just discussed, discrete descriptions are computationally expensive because of the large number of rods that must be considered in order to derive sufficiently accurate model outputs. On the other hand, Fokker–Planck-based descriptions are rarely considered, precisely because of the curse of dimensionality that the introduction of conformation coordinates (the rod's orientation in the case considered here) implies. Thus, standard mesh-based discretization techniques, such as finite differences, finite elements or finite volumes, fail when addressing models defined in high-dimensional spaces.

For these reasons, mesoscopic models were coarsened to derive macroscopic models defined in standard physical domains, involving space and time. At the macroscopic scale, the orientation distribution function is substituted with its moments for describing the microstructure. Usually macroscopic descriptions of rod suspensions are based on the use of the first two non-vanishing moments, the second, **a**, and the fourth, **A**, order moments (the odd moments vanish because of the symmetry of the pdf), with the former being given by

$$\mathbf{a} = \int_{\mathscr{S}} \mathbf{p} \otimes \mathbf{p}\, \Psi(\mathbf{p})\, d\mathbf{p}, \tag{4.61}$$

and the latter by Eq. (4.58).

Macroscopic Scale: Microstructural Evolution

The microstructural evolution described at the macroscopic scale considers the time evolution of the pdf moments. The time evolution of the second order orientation tensor, when replacing $\dot{\Psi}$ with its expression given by the Fokker–Planck equation (4.69), and after integrating by parts, results in

$$\begin{aligned}
\dot{\mathbf{a}} &= \int_{\mathscr{S}} (\dot{\mathbf{p}} \otimes \mathbf{p} + \mathbf{p} \otimes \dot{\mathbf{p}})\ \Psi\, d\mathbf{p} \\
&= \int_{\mathscr{S}} (\nabla \mathbf{v} \cdot \mathbf{p} - (\nabla \mathbf{v} : (\mathbf{p} \otimes \mathbf{p}))\ \mathbf{p}) \otimes \mathbf{p}\, \Psi\, d\mathbf{p} \\
&+ \int_{\mathscr{S}} \mathbf{p} \otimes (\nabla \mathbf{v} \cdot \mathbf{p} - (\nabla \mathbf{v} : (\mathbf{p} \otimes \mathbf{p}))\ \mathbf{p})\ \Psi\, d\mathbf{p} \\
&= \nabla \mathbf{v} \cdot \mathbf{a} + \mathbf{a} \cdot (\nabla \mathbf{v})^T - 2\, \mathbf{A} : \nabla \mathbf{v},
\end{aligned} \tag{4.62}$$

which depends on the fourth order moment **A**. The time derivative of the fourth order moment, using the same rationale, implies the sixth order one $\mathscr{A}$, and so on.

Another possibility for deducing Eq. (4.62) lies in applying the rationale followed for deriving the hydrodynamic equations in Sect. 4.2.1. Thus, we multiply the Fokker–Planck equation (4.69) by $\mathbf{p} \otimes \mathbf{p}$ and integrate on the unit ball surface $\mathscr{S}$, making use of the integration by parts for the term involving $\nabla_p(\dot{p}\Psi)$. This procedure will be considered later when addressing polymeric systems.

Thus, if we consider the microstructure described from '**a**,' a closure relation is needed in order to express the fourth order moment **A** as a function of the lower order moments ($\mathbf{a}$ in the present case). Different closure relations have been introduced and widely used [5]. When considering, for example, the quadratic closure relation (which is only exact when all the rods are locally aligned in the same direction), the fourth order moment results in

$$\mathbf{A} \approx \mathbf{a} \otimes \mathbf{a}, \tag{4.63}$$

which allows us to write

$$\dot{\mathbf{a}} \approx \nabla \mathbf{v} \cdot \mathbf{a} + \mathbf{a} \cdot (\nabla \mathbf{v})^T - 2\,(\nabla \mathbf{v} : \mathbf{a})\,\mathbf{a}, \tag{4.64}$$

or, invoking symmetry considerations again

$$\dot{\mathbf{a}} \approx \nabla \mathbf{v} \cdot \mathbf{a} + \mathbf{a} \cdot (\nabla \mathbf{v})^T - 2\,(\mathbf{D} : \mathbf{a})\,\mathbf{a}. \tag{4.65}$$

Macroscopic Scale: Moment-Based Stress

We previously obtained the expression of the rod's population contribution to the stress

$$\boldsymbol{\tau} = 2\eta N_p(\mathbf{A} : \nabla \mathbf{v}), \tag{4.66}$$

which implies the use of the fourth order moment $\mathbf{A}$. When $\mathbf{A}$ is calculated from the pdf Ψ by using (4.58) within the mesoscopic framework, there are no closure issues. However, when one proceeds at the macroscopic scale in which the pdf is no longer available, a closure relation must be considered for either

- writing $\mathbf{A}$ from the knowledge of $\mathbf{a}$, the latter calculated by integrating (4.62) with an appropriate closure relation (e.g., Eq. (4.64) when considering the quadratic closure),

or

- calculating $\mathbf{A}$ by solving the equation that governs its time evolution in which, as just mentioned the sixth order moment appears, again requiring an appropriate closure.

The first route is the simplest one and the most used in practice, which leads to

$$\boldsymbol{\tau} = 2\eta N_p(\mathbf{A}^{cr}(\mathbf{a}) : \nabla \mathbf{v}), \tag{4.67}$$

where the superscript '*cr*" refers to the use of an appropriate closure, relationship.

When considering the quadric closure, the stress writes:

$$\boldsymbol{\tau} = 2\eta N_p(\mathbf{a} : \nabla \mathbf{v})\mathbf{a} = 2\eta N_p(\mathbf{a} : \mathbf{D})\mathbf{a}. \tag{4.68}$$

4.3.2 *On the Solution of the Fokker–Planck Equation*

As discussed in the previous section, the solution of the Fokker–Planck equation seems to be a real issue when using well experienced mesh-based discretization techniques, due to the curse of dimensionality that it implies. As this issue is a recurrent difficulty in all the models treated in this chapter, before moving on to the description of other kinetic theory models, we will briefly discuss different simulation alternatives.

Since kinetic theory descriptions involve a probability distribution function depending on space, time and a number of conformational coordinates, the associated Fokker–Planck equations suffer the so-called curse of dimensionality typical of problems defined in highly dimensional spaces.

Thus, mesh-based discretization techniques fail for discretizing the problem, because the number of degrees of freedom involved in a mesh or grid increases exponentially with the space dimension.

In what follows, we consider some alternatives to standard mesh-based discretization that are able to address the solution of Fokker–Planck equations associated with kinetic theory descriptions.

We consider, without loss of generality, a generic Fokker–Planck equation involving the pdf $\Psi(\mathbf{x}, t, \mathbf{a})$, where here $\mathbf{a}$ represents a set of arbitrary conformational coordinates (it is of fundamental importance not to confuse conformational coordinates $\mathbf{a}$ and the orientation tensor considered in the previous section) and diffusion terms are represented by fluxes $\mathbf{q}_x$ and $\mathbf{q}_a$ operating, respectively, in the physical and conformational spaces:

$$\frac{\partial \Psi}{\partial t} + \nabla_x \cdot (\mathbf{v}\, \Psi) + \nabla_a \cdot (\dot{\mathbf{a}}\, \Psi) = -\nabla_x \cdot \mathbf{q}_x - \nabla_a \cdot \mathbf{q}_a. \tag{4.69}$$

4.3.2.1 Method of Particles for Solving Advection-Dominated Problems

This technique, described in detail in [6, 7], consists in approximating the initial distribution $\Psi(\mathbf{x}, t = 0, \mathbf{a})$ from $\mathscr{M}$ Dirac's masses $\mathbf{a}_i^0$ at each one of the $\mathscr{Q}$ positions $\mathbf{x}_j^0$:

$$\Psi(\mathbf{x}, t = 0, \mathbf{a}) = \sum_{j=1}^{\mathscr{Q}} \sum_{i=1}^{\mathscr{M}} \alpha_i^j \, \delta(\mathbf{a} - \mathbf{a}_i^0)\, \delta(\mathbf{x} - \mathbf{x}_j^0). \tag{4.70}$$

This represents a sort of approximation based on $\mathscr{Q} \cdot \mathscr{M}$ computational particles $\mathscr{P}_{ij}$ with initial positions and conformations given by

$$\begin{cases} \mathbf{x}_{ij}^0 = \mathbf{x}_j^0, & i = 1, \ldots, \mathscr{M}; \quad j = 1, \ldots, \mathscr{Q} \\ \mathbf{a}_{ij}^0 = \mathbf{a}_i^0, & i = 1, \ldots, \mathscr{M}; \quad j = 1, \ldots, \mathscr{Q} \end{cases}, \tag{4.71}$$

and whose position and conformation will be evaluated all along the flow simulation, from which the distribution will be reconstructed.

When considering the purely advective balance equation

$$\frac{\partial \Psi}{\partial t} + \nabla_x \cdot (\mathbf{v}\, \Psi) + \nabla_a \cdot (\dot{\mathbf{a}}\, \Psi) = 0, \tag{4.72}$$

the time evolution of position and conformation of each particle $\mathscr{P}_{ij}$ is calculated by integrating

$$\begin{cases} \mathbf{x}_{ij}(t) = \mathbf{x}_{ij}^0 + \int_{\tau=0}^{\tau=t} \mathbf{v}(\mathbf{x}_{ij}(\tau))\, d\tau \\ \mathbf{a}_{ij}(t) = \mathbf{a}_{ij}^0 + \int_{\tau=0}^{\tau=t} \dot{\mathbf{a}}_{ij}(\mathbf{a}_{ij}(\tau), \mathbf{x}_{ij}(\tau))\, d\tau \end{cases}. \tag{4.73}$$

As the position update only depends on the velocity field, which itself only depends on the position, it can be stressed that particles $\mathscr{P}_{ij}, i = 1, \ldots, \mathscr{M}$ are following the same trajectory in the physical space, having $\mathbf{x}_j^0$ as their departure point.

Now, the orientation distribution at time t can be reconstructed from

$$\Psi(\mathbf{x}, t, \mathbf{a}) = \sum_{j=1}^{\mathscr{Q}} \sum_{i=1}^{\mathscr{M}} \alpha_i^j\, \delta(\mathbf{a} - \mathbf{a}_{ij}(t))\, \delta(\mathbf{x} - \mathbf{x}_{ij}(t)). \tag{4.74}$$

Obviously, smoother representations can be obtained by considering appropriate regularizations of Dirac's distribution, as the one usually performed within the SPH (Smooth Particles Hydrodynamics) framework [6, 8].

When considering diffusion terms, there are two main routes based on the use of particles, one of stochastic nature, the other fully deterministic.

To illustrate the procedure in cases in which models involve diffusion terms, we consider the Fokker–Planck equation

$$\frac{\partial \Psi}{\partial t} + \nabla_x \cdot (\mathbf{v}\, \Psi) + \nabla_a \cdot (\dot{\mathbf{a}}\, \Psi) = -\nabla_x \cdot \mathbf{q}_x - \nabla_a \cdot \mathbf{q}_a, \tag{4.75}$$

where $\mathbf{q}_x$ and $\mathbf{q}_a$ are two diffusive fluxes operating in the physical and conformational spaces, respectively, both modeled on a Fick-type law:

$$\begin{cases} \mathbf{q}_x = -\mathbf{D}_x \cdot \nabla_x \Psi \\ \mathbf{q}_a = -\mathbf{D}_a \cdot \nabla_a \Psi \end{cases}. \tag{4.76}$$

- Within the stochastic framework, diffusion terms can be modeled from appropriate random variables within a Lagrangian or a Eulerian description, the latter one being known as Brownian Configurations Fields (BCF). Both approaches were considered in our former works on the solution of Fokker–Planck equations [7, 9]. Within the Lagrangian stochastic framework, and starting from the initial cloud of computational particles $\mathscr{P}_{ij}$ representing the initial distribution $\Psi(\mathbf{x}, t = 0, \mathbf{a})$, the simplest particles updating reads as

$$\begin{cases} \mathbf{x}_{ij}(t_{n+1}) = \mathbf{x}_{ij}(t_n) + \mathbf{v}(\mathbf{x}_{ij}(t_n))\ \Delta t + \mathscr{R}_x(\Delta t) \\ \mathbf{a}_{ij}(t_{n+1}) = \mathbf{a}_{ij}(t_n) + \dot{\mathbf{a}}_{ij}(\mathbf{a}_{ij}(t_n), \mathbf{x}_{ij}(t_n))\ \Delta t + \mathscr{R}_a(\Delta t) \end{cases}, \tag{4.77}$$

where Δt is the time step and both random updates $\mathscr{R}_x$ and $\mathscr{R}_a$ depend on the chosen time step (see [10] for more details, as well as for advanced stochastic integrations).

Obviously, because of the random effects operating in the physical space, the $\mathscr{M}$ particles initially located at each position $\mathbf{x}_j^0$, $j = 1, \ldots, \mathscr{Q}$, will follow different trajectories in the physical space along the simulation. In order to obtain sufficiently accurate results, we must consider a sufficiently rich representation, that is, a large population of particles. For this purpose, we must consider sufficiently large $\mathscr{M}$ and $\mathscr{Q}$. The large number of particles to be tracked seems a disadvantage of the approach at first sight, but it must be noted that the integration of each particle is completely independent of all the others, making possible the use of HPC on massively parallel computing platforms.

- The technique introduced to treat purely advective equations can be extended to consider diffusion contributions, as was described in [8] and that we revisit in what follows, within a fully deterministic approach. Equation (4.75) can be rewritten as:

$$\frac{\partial \Psi}{\partial t} + \nabla_x \cdot \left(\left(\mathbf{v} + \frac{\mathbf{q}_x}{\Psi} \right) \Psi \right) + \nabla_a \cdot \left(\left((\dot{\mathbf{a}} + \frac{\mathbf{q}_a}{\Psi} \right) \Psi \right) = 0 \tag{4.78}$$

or

$$\frac{\partial \Psi}{\partial t} + \nabla_x \cdot (\tilde{\mathbf{v}}\ \Psi) + \nabla_a \cdot \left(\tilde{\dot{\mathbf{a}}}\ \Psi \right) = 0, \tag{4.79}$$

where the effective velocities $\tilde{\mathbf{v}}$ and $\tilde{\dot{\mathbf{a}}}$ are given by:

$$\begin{cases} \tilde{\mathbf{v}} = \mathbf{v} - \frac{1}{\Psi}\, \mathbf{D}_x \cdot \nabla_x \Psi \\ \tilde{\dot{\mathbf{a}}} = \dot{\mathbf{a}} - \frac{1}{\Psi}\, \mathbf{D}_a \cdot \nabla_a \Psi \end{cases}. \tag{4.80}$$

Now, the integration scheme (4.73) can be applied by replacing material and conformation velocities, $\mathbf{v}$ and $\dot{\mathbf{a}}$, with their effective counterparts $\tilde{\mathbf{v}}$ and $\tilde{\dot{\mathbf{a}}}$:

$$\begin{cases} \mathbf{x}_{ij}(t) = \mathbf{x}_{ij}^0 + \int_{\tau=0}^{\tau=t} \tilde{\mathbf{v}}(\mathbf{x}_{ij}(\tau))\ d\tau \\ \mathbf{a}_{ij}(t) = \mathbf{a}_{ij}^0 + \int_{\tau=0}^{\tau=t} \tilde{\dot{\mathbf{a}}}_{ij}(\mathbf{a}_{ij}(\tau), \mathbf{x}_{ij}(\tau))\ d\tau \end{cases}. \tag{4.81}$$

This fully deterministic particle description requires far fewer particles than its stochastic counterpart, but as noted in Eq. (4.80), the calculation of the effective material and conformational velocities requires the derivative of the pdf Ψ with respect to both the physical and conformational coordinates. To achive this, the distribution must be reconstructed all along the simulation (at each time step), which constitutes a serious drawback for its implementation on massively parallel computing platforms. Moreover, to make possible the calculation of the distribution derivatives, the Dirac distribution must be regularized in order to ensure its derivability.

4.3.2.2 Separate Representations for Solving Difusion-Dominated Problems

When the diffusion effects are dominant, the techniques presented in the previous section become inefficient, because they require an excessive number of particles to produce sufficiently accurate results, in particular, for reconstructing the distribution. In this case, standard mesh-based discretizations seem a better choice. However, as discussed before, mesh-based discretizations fail when addressing highly dimensional models as is the case when addressing the solution of the previously introduced Fokker–Planck equation. Separate representations seem the most appealing choice.

Considering the Fokker–Planck equation

$$\frac{\partial \Psi}{\partial t} + \nabla_x \cdot (\mathbf{v}\, \Psi) + \nabla_a \cdot (\dot{\mathbf{a}}\, \Psi) = \nabla_x \left(\mathbf{D}_x \cdot \nabla_x \Psi\right) + \nabla_a \left(\mathbf{D}_a \cdot \nabla_a \Psi\right), \qquad (4.82)$$

there are many separate representation choices. The most natural one consists in separating time, physical and conformational spaces, i.e.,

$$\Psi(\mathbf{x}, t, \mathbf{a}) \approx \sum_{i=1}^{N} X_i(\mathbf{x}) \cdot T_i(t) \cdot A_i(\mathbf{a}). \qquad (4.83)$$

Thus, when proceeding with the Proper Generalized Decomposition – PGD – constructor [11–14], we must solve on the order of N 2D or 3D (depending on the dimension of the physical space) boundary value problems – BVP – for calculating functions $X_i(\mathbf{x})$, the same number of 1D initial value problems – IVP – for calculating functions $T_i(t)$, and finally, the same number of problems involving the conformational coordinates for calculating functions $A_i(\mathbf{a})$.

4.3.3 Dilute Suspensions of Brownian Rods

In Sect. 4.3.1, Brownian effects were neglected. The Brownian effects are the result of the fluid molecules bombardment acting on the rod beads. This section focuses on the modeling of such effects, which are particularly important when considering nano-charges (e.g., carbon nanotubes – CNT). Moreover, as that bombardment results in a diffusion mechanism, inducing a sort of randomizing effect, the same modeling framework is usually retained to address semi-dilute or semi-concentrated suspensions of rods in order to take into account the inter-rod interactions.

4.3.3.1 Microscopic Description

In this case, the rod beads are subjected to the hydrodynamic forces and the ones coming from the bombardment. The first one was introduced previously,

$$\mathbf{F}^H = \xi L \left(\nabla \mathbf{v} \cdot \mathbf{p} - \dot{\mathbf{p}} \right), \tag{4.84}$$

where the superscript 'H" refers to its hydrodynamic nature. Now, the Brownian force $\mathbf{F}^B$ is assumed to act during a short time interval δt following a certain statistical distribution concerning its magnitude and its orientation. The first one, by virtue of the central limit theorem, is assumed to be described by a Gaussian distribution of zero mean and a certain standard deviation and the one related to the orientation by a uniform distribution on the unit circle $\mathscr{C}$ (in 2D) or in the unit sphere $\mathscr{S}$ (in 3D).

Brownian forces acting in the direction of the rod are assumed to be equilibrated. However, the components of those forces perpendicular to the rods axis contribute to the rod's rotation, and thus they affect its rotary velocity. In what follows, for the sake of clarity, we restrict our analysis to the 2D case. For inertialess rods, the resultant moment must vanish, and consequently

$$\mathbf{F}^H \cdot \mathbf{t} + \mathbf{F}^B \cdot \mathbf{t} = 0, \tag{4.85}$$

$\mathbf{t}$ being the unit vector tangent to the unit circle defined by $\mathbf{t} = \frac{\dot{\mathbf{p}}}{\|\dot{\mathbf{p}}\|}$.

By injecting the hydrodynamic force (4.84) into (4.85), it results that

$$\mathbf{t}^T \cdot \nabla \mathbf{v} \cdot \mathbf{p} - \|\dot{\mathbf{p}}\| = -\frac{\mathbf{F}^B \cdot \mathbf{t}}{\xi L}, \tag{4.86}$$

and multiplying by $\mathbf{t}$, taking into account that $\dot{\mathbf{p}} = \|\dot{\mathbf{p}}\|\mathbf{t}$, it results that

$$\dot{\mathbf{p}} = \left(\mathbf{t}^T \cdot \nabla \mathbf{v} \cdot \mathbf{p} \right) \mathbf{t} + \frac{\mathbf{F}^B \cdot \mathbf{t}}{\xi L} \mathbf{t}, \tag{4.87}$$

which, using the vectorial equivalence

$$\left(\mathbf{t}^T \cdot \nabla \mathbf{v} \cdot \mathbf{p} \right) \mathbf{t} = \nabla \mathbf{v} \cdot \mathbf{p} - \left(\mathbf{p}^T \cdot \nabla \mathbf{v} \cdot \mathbf{p} \right) \mathbf{p}, \tag{4.88}$$

results in

$$\dot{\mathbf{p}} = \nabla \mathbf{v} \cdot \mathbf{p} - \left(\mathbf{p}^T \cdot \nabla \mathbf{v} \cdot \mathbf{p} \right) \mathbf{p} + \frac{\mathbf{F}^B \cdot \mathbf{t}}{\xi L} \mathbf{t}$$

$$= \nabla \mathbf{v} \cdot \mathbf{p} - \left(\mathbf{p}^T \cdot \nabla \mathbf{v} \cdot \mathbf{p}\right) \mathbf{p} + \frac{\mathbf{F}^B - (\mathbf{F}^B \cdot \mathbf{p})\mathbf{p}}{\xi L}, \tag{4.89}$$

where we can notice that the rotary velocity is given by the Jeffery expression $\dot{\mathbf{p}}^J$ complemented by a term accounting for the Bownian effects $\dot{\mathbf{p}}^B$,

$$\dot{\mathbf{p}}^J = \nabla \mathbf{v} \cdot \mathbf{p} - \left(\mathbf{p}^T \cdot \nabla \mathbf{v} \cdot \mathbf{p}\right) \mathbf{p}, \tag{4.90}$$

and

$$\dot{\mathbf{p}}^B = \frac{\mathbf{F}^B - (\mathbf{F}^B \cdot \mathbf{p})\mathbf{p}}{\xi L}, \tag{4.91}$$

from which

$$\dot{\mathbf{p}} = \dot{\mathbf{p}}^J + \dot{\mathbf{p}}^B. \tag{4.92}$$

Now, we discuss the effects of such a Brownian contribution to the extra-stress tensor.

Consider a rod aligned along the x-axis, such that $\mathbf{p}^T = (1, 0)$ and the fluid is at rest. This rod undergoes a continuous bombardment from the suspending fluid molecules. The component of forces aligned with the rod axis does not contribute to the rod's rotation, and by averaging it in $\Delta t \gg \delta t$ (Δt and δt being the rod kinematics' characteristic time and the one related to the bombardement, respectively), the resulting contribution vanishes. Contrastingly, the component perpendicular to the rod will contribute to the stress. To derive the expression of this Brownian contribution, we consider that, due to an impact, the rod rotates by a small angle, e.g., $\delta\theta > 0$, with the rod's orientation defined by $\mathbf{p}_{\delta\theta}$. We consider the unit tangent vector $\mathbf{t}_{\delta\theta}$ such that $\mathbf{p}_{\delta\theta} \times \mathbf{t}_{\delta\theta} = \mathbf{e}_z$ ($\mathbf{e}_z$ is the unit vector in the out-of-plane direction in the present 2D analysis). Considering the Brownian force applied at that position, i.e., $\|\mathbf{F}^B\| \cdot \mathbf{t}_{\delta\theta}$, it results in the contribution to the virial stress

$$-\|\mathbf{F}^B\| \, \mathbf{p}_{\delta\theta} \otimes \mathbf{t}_{\delta\theta} = \|\mathbf{F}^B\| \begin{pmatrix} \sin(\delta\theta)\,\cos(\delta\theta) & -\cos^2(\delta\theta) \\ \sin^2(\delta\theta) & -\sin(\delta\theta)\,\cos(\delta\theta) \end{pmatrix}, \tag{4.93}$$

where the negative sign accounts for the fact that the hydrodynamic force applies in the opposite direction of the Brownian force.

We can notice two facts in that expression: (i) the trace is zero, and (ii) the contribution is non-symmetric. However, we can imagine that a little bit later, a Brownian force will apply in the opposite direction. Invoking ergodicity (allowing us to replace time by ensemble averages), we could assume that another particle initially aligned along $\mathbf{p}$ receives the opposite impact, leading to an angle $-\delta\theta$, from which

$$\|\mathbf{F}^B\|\, \mathbf{p}_{-\delta\theta} \otimes \mathbf{t}_{-\delta\theta} = \|\mathbf{F}^B\| \begin{pmatrix} \sin(\delta\theta)\,\cos(\delta\theta) & \cos^2(\delta\theta) \\ -\sin^2(\delta\theta) & -\sin(\delta\theta)\,\cos(\delta\theta) \end{pmatrix}, \tag{4.94}$$

and then, by averaging, it gives a Brownian contribution to the extra-stress associated with the x-direction ($\varphi = 0$)

$$\boldsymbol{\tau}^B_{\varphi=0} = \begin{pmatrix} \beta & 0 \\ 0 & -\beta \end{pmatrix} = \beta \begin{pmatrix} 1 & 0 \\ 0 & -1 \end{pmatrix} = \beta\,\mathbf{U}, \tag{4.95}$$

which becomes symmetric and traceless.

Now, for rods aligned in any other direction φ, it suffices to apply a rotation of angle φ to the tensor $\boldsymbol{\tau}^B_{\varphi=0}$ according to

$$\boldsymbol{\tau}^B_{\varphi} = \beta\, \mathbf{R}^T_{\varphi} \cdot \mathbf{U} \cdot \mathbf{R}_{\varphi}, \tag{4.96}$$

with

$$\mathbf{R}_{\varphi} = \begin{pmatrix} \cos\varphi & \sin\varphi \\ -\sin\varphi & \cos\varphi \end{pmatrix}, \tag{4.97}$$

which ultimately yields

$$\boldsymbol{\tau}^B_{\varphi} = 2\beta \left(\begin{pmatrix} \cos^2\varphi & \sin\varphi\,\cos\varphi \\ \sin\varphi\,\cos\varphi & \sin^2\varphi \end{pmatrix} - \begin{pmatrix} \frac{1}{2} & 0 \\ 0 & \frac{1}{2} \end{pmatrix} \right), \tag{4.98}$$

which can be rewritten as

$$\boldsymbol{\tau}^B_{\mathbf{p}_\varphi} = 2\beta \left(\mathbf{p}_\varphi \otimes \mathbf{p}_\varphi - \frac{\mathbf{I}}{2} \right), \tag{4.99}$$

where $\mathbf{I}$ is the identity tensor and $\mathbf{p}_\varphi$ the unit vector defined by the angle φ.

For a population of rods $\mathbf{p}_i$, $i = 1, \ldots, N$, the contribution of Brownian effects is ultimately given by

$$\boldsymbol{\tau}^B = 2\beta \sum_{i=1}^{N} \left(\mathbf{p}_i \otimes \mathbf{p}_i - \frac{\mathbf{I}}{2} \right). \tag{4.100}$$

4.3.3.2 Mesoscopic Description

The mesoscopic discrete description can be summarized as follows:

- Given the flow velocity field $\mathbf{v}(\mathbf{x}, t)$, the initial position of the centre of gravity of each rod $\mathbf{x}_i^G(t = 0)$ and its orientation $\mathbf{p}_i(t = 0)$, or knowing both them at time t;
- For each rod $i = 1, \ldots, N$:

– Update the position of the centre of gravity by integrating:

$$\frac{d\mathbf{x}_i^G}{dt} = \mathbf{v}(\mathbf{x}_i^G, t). \tag{4.101}$$

– Update its orientation by integrating

$$\frac{d\mathbf{p}_i}{dt} = \nabla\mathbf{v}|_{\mathbf{x}_i^G,t} \cdot \mathbf{p}_i - \left(\mathbf{p}_i^T \cdot \nabla\mathbf{v}|_{\mathbf{x}_i^G,t} \cdot \mathbf{p}_i\right) \mathbf{p}_i + \frac{\mathbf{F}_i^B - (\mathbf{F}_i^B \cdot \mathbf{p}_i)\,\mathbf{p}_i}{\xi L}. \tag{4.102}$$

- Compute the stress at position $\mathbf{x}$ and time t by considering all the rods inside a control volume $\mathscr{V}(\mathbf{x})$ centered at $\mathbf{x}$ by applying

$$\boldsymbol{\tau}(\mathbf{x}, t) = 2\eta\, \mathbf{D}(\mathbf{x}, t)$$
$$+ \sum_{i \in \mathscr{V}(\mathbf{x})} \left(2\eta N_p\, \nabla\mathbf{v}|_{\mathbf{x},t} : (\mathbf{p}_i \otimes \mathbf{p}_i \otimes \mathbf{p}_i \otimes \mathbf{p}_i) + 2\beta \left(\mathbf{p}_i \otimes \mathbf{p}_i - \frac{\mathbf{I}}{2}\right)\right). \tag{4.103}$$

In the case of coupled models, this stress will serve to update the velocity field, however, as previously indicated, in the present text, we consider the velocity field unperturbed by the presence and orientation of the rods (decoupled modeling).

On the other hand, the mesoscopic continuous description considers a diffusion term in the Fokker–Planck equation

$$\frac{\partial \Psi}{\partial t} + \nabla_x \cdot (\mathbf{v}\Psi) + \nabla_p \cdot (\dot{\mathbf{p}}\Psi) = D_r \nabla_p^2 \Psi, \tag{4.104}$$

with the flow-induced orientation $\dot{\mathbf{p}}$ given by Jeffery's equation (4.49), and where D_r is the so-called rotary diffusion.

We can not that in absence of flow, i.e., $\mathbf{v}(\mathbf{x}, t) = \mathbf{0}$, the Fokker–Planck equation reduces to

$$\frac{\partial \Psi}{\partial t} = D_r \nabla_p^2 \Psi, \tag{4.105}$$

which ensures a steady state consisting of an isotropic orientation distribution, i.e., $\Psi(\mathbf{x}, t \to \infty, \mathbf{p}) = \frac{1}{2\pi}$ in 2D and $\Psi(\mathbf{x}, t \to \infty, \mathbf{p}) = \frac{1}{4\pi}$ in 3D. The higher the rotational diffusion, the faster the isotropic orientation distribution is reached.

Thus, at the continuous mesoscopic level, the introduction of Brownian effects seems quite simple. The question is: what are the discrete and macroscopic counterparts of the just introduced diffusion term?

The Fokker–Planck equation can be rewritten in the form

$$\frac{\partial \Psi}{\partial t} + \nabla_x \cdot (\mathbf{v}\Psi) + \nabla_p \cdot (\dot{\mathbf{p}}\Psi) - D_r \nabla_p^2 \Psi$$

$$= \frac{\partial \Psi}{\partial t} + \nabla_x \cdot (\mathbf{v}\Psi) + \nabla_p \cdot \left(\dot{\tilde{\mathbf{p}}}\Psi\right) = 0, \tag{4.106}$$

where the effective rotational velocity $\dot{\tilde{\mathbf{p}}}$ is given by

$$\dot{\tilde{\mathbf{p}}} = \nabla\mathbf{v} \cdot \mathbf{p} - \left(\mathbf{p}^T \cdot \nabla\mathbf{v} \cdot \mathbf{p}\right)\mathbf{p} - D_r \frac{\nabla_p \Psi}{\Psi}, \tag{4.107}$$

which contains the flow-induced Jeffery contribution $\dot{\mathbf{p}}^J$ plus the Brownian one $\dot{\mathbf{p}}^B$, i.e., $\dot{\tilde{\mathbf{p}}} = \dot{\mathbf{p}}^J + \dot{\mathbf{p}}^B$ with

$$\dot{\mathbf{p}}^J = \nabla\mathbf{v} \cdot \mathbf{p} - \left(\mathbf{p}^T \cdot \nabla\mathbf{v} \cdot \mathbf{p}\right)\mathbf{p}, \tag{4.108}$$

and

$$\dot{\mathbf{p}}^B = -D_r \frac{\nabla_p \Psi}{\Psi}. \tag{4.109}$$

Within the mesoscopic continuous description, the Brownian contribution to the extra-stress tensor results from the generalization of Eq. (4.100) by substituting the sum with that corresponding integral

$$\boldsymbol{\tau}^B = 2\tilde{\beta} \int_{\mathcal{S}} \left(\mathbf{p} \otimes \mathbf{p} - \frac{\mathbf{I}}{2}\right) \Psi(\mathbf{p})\, d\mathbf{p} = 2\tilde{\beta}\left(\mathbf{a} - \frac{\mathbf{I}}{2}\right). \tag{4.110}$$

In summary, the mesoscopic continuous description consists of:

- Given the flow velocity field $\mathbf{v}(\mathbf{x}, t)$ and the initial orientation distribution $\Psi(\mathbf{x}, t = 0, \mathbf{p})$
- Solve the Fokker–Planck equation that governs the rod's orientation distribution:

$$\frac{\partial \Psi}{\partial t} + \nabla_x \cdot (\mathbf{v}\Psi) + \nabla_p \cdot (\dot{\mathbf{p}}\Psi) = D_r \nabla_p^2 \Psi, \tag{4.111}$$

with

$$\dot{\mathbf{p}} = \nabla\mathbf{v} \cdot \mathbf{p} - \left(\mathbf{p}^T \cdot \nabla\mathbf{v} \cdot \mathbf{p}\right)\mathbf{p}. \tag{4.112}$$

- Compute the stress tensor:

$$\boldsymbol{\tau} = \boldsymbol{\tau}^f + \boldsymbol{\tau}^r = 2\eta\,\mathbf{D} + 2\eta N_p\,(\mathbf{D} : \mathbf{A}) + 2\tilde{\beta}\left(\mathbf{a} - \frac{\mathbf{I}}{2}\right). \tag{4.113}$$

4.3.3.3 Macroscopic Description

When moving towards the macroscopic scale, the Brownian contribution to the extra-stress is defined by Eq. (4.110), however, at the macroscopic scale, the microstructure is defined by the different moments of the orientation distribution. In what follows, we will derive the contribution of Brownian effects to the equation governing the evolution of the second order moment.

We start from the second order moment definition

$$\mathbf{a} = \int_{\mathscr{S}} \mathbf{p} \otimes \mathbf{p} \, \Psi \, d\mathbf{p}, \tag{4.114}$$

whose time derivative now involves the effective rotational velocity $\dot{\tilde{\mathbf{p}}}$

$$\dot{\mathbf{a}} = \int_{\mathscr{S}} \left(\dot{\tilde{\mathbf{p}}} \otimes \mathbf{p} + \mathbf{p} \otimes \dot{\tilde{\mathbf{p}}} \right) \Psi \, d\mathbf{p}, \tag{4.115}$$

As proved in the previous sections, the effective rotational velocity contains the flow-induced contribution given by the Jeffery expression $\dot{\mathbf{p}}^J$ and the one induced by the Brownian effects $\dot{\mathbf{p}}^B$, given by Eqs. (4.108) and (4.109), respectively. With this decomposition, Eq. (4.115) can be written as

$$\begin{aligned}\dot{\mathbf{a}} &= \int_{\mathscr{S}} \left((\dot{\mathbf{p}}^J + \dot{\mathbf{p}}^B) \otimes \mathbf{p} + \mathbf{p} \otimes (\dot{\mathbf{p}}^J + \dot{\mathbf{p}}^B) \right) \Psi \, d\mathbf{p} \\ &= \int_{\mathscr{S}} \left(\dot{\mathbf{p}}^J \otimes \mathbf{p} + \mathbf{p} \otimes \dot{\mathbf{p}}^J \right) \Psi \, d\mathbf{p} \\ &+ \int_{\mathscr{S}} \left(\dot{\mathbf{p}}^B \otimes \mathbf{p} + \mathbf{p} \otimes \dot{\mathbf{p}}^B \right) \Psi \, d\mathbf{p} = \dot{\mathbf{a}}^J + \dot{\mathbf{a}}^B,\end{aligned} \tag{4.116}$$

where the flow induced microstructure evolution $\dot{\mathbf{a}}^J$ is given by

$$\dot{\mathbf{a}}^J = \nabla \mathbf{v} \cdot \mathbf{a} + \mathbf{a} \cdot (\nabla \mathbf{v})^T - 2\mathbf{A} : \mathbf{D}. \tag{4.117}$$

We now calculate the expression of the remaining contribution $\dot{\mathbf{a}}^B$

$$\dot{\mathbf{a}}^B = \int_{\mathscr{S}} \left(\dot{\mathbf{p}}^B \otimes \mathbf{p} + \mathbf{p} \otimes \dot{\mathbf{p}}^B \right) \Psi \, d\mathbf{p}, \tag{4.118}$$

with $\dot{\mathbf{p}}^B$ given by

$$\dot{\mathbf{p}}^B = -D_r \frac{\nabla_p \Psi}{\Psi}. \tag{4.119}$$

For the sake of clarity, we again consider the 2D case that allows us to write

$$\dot{\mathbf{p}}^B = -D_r \frac{\|\nabla_p \Psi\|}{\Psi} \, \mathbf{t}, \tag{4.120}$$

with $\mathbf{t}$ being the unit tangent vector to the unit circle aligned with the rotary velocity. In this case, Eq. (4.118) reduces to

$$\dot{\mathbf{a}}^B = -D_r \int_{\mathscr{S}} (\mathbf{t} \otimes \mathbf{p} + \mathbf{p} \otimes \mathbf{t}) \, \frac{\partial \Psi}{\partial \theta} \, d\theta. \tag{4.121}$$

Now, integrating Eq. (4.121) by parts and taking into account

$$\frac{d\mathbf{p}}{d\theta} = \mathbf{t} \tag{4.122}$$

and

$$\frac{d\mathbf{t}}{d\theta} = -\mathbf{p}, \tag{4.123}$$

it results that

$$\dot{\mathbf{a}}^B = -2D_r \int_{\mathscr{S}} (\mathbf{p} \otimes \mathbf{p} - \mathbf{t} \otimes \mathbf{t}) \, \Psi(\theta) \, d\theta. \tag{4.124}$$

It is easy to prove that

$$\mathbf{t} \otimes \mathbf{t} + \mathbf{p} \otimes \mathbf{p} = \mathbf{I} \;\rightarrow\; \mathbf{t} \otimes \mathbf{t} = \mathbf{I} - \mathbf{p} \otimes \mathbf{p}, \tag{4.125}$$

which allows us to write Eq. (4.124) in the form

$$\dot{\mathbf{a}}^B = -2D_r \int_{\mathscr{S}} (2(\mathbf{p} \otimes \mathbf{p}) - \mathbf{I}) \, \Psi(\theta) \, d\theta = -2D_r \, (2\mathbf{a} - \mathbf{I}) \tag{4.126}$$

or

$$\dot{\mathbf{a}}^B = -4D_r \left(\mathbf{a} - \frac{\mathbf{I}}{2} \right). \tag{4.127}$$

When considering a 3D situation, the resulting expression is given by

$$\dot{\mathbf{a}}^B = -6D_r \left(\mathbf{a} - \frac{\mathbf{I}}{3} \right). \tag{4.128}$$

We can notice that in the 2D case, and in the absence of flow, $\dot{\mathbf{a}}^J = \mathbf{0}$, and thus $\dot{\mathbf{a}} = \dot{\mathbf{a}}^B$

$$\dot{\mathbf{a}} = -4D_r \left(\mathbf{a} - \frac{\mathbf{I}}{2} \right), \tag{4.129}$$

ensuring an isotropic steady state, i.e., $\mathbf{a}(t \rightarrow \infty) = \frac{\mathbf{I}}{2}$.

In summary, the decoupled macroscopic description consists of:

- Given the flow velocity field $\mathbf{v}(\mathbf{x}, t)$ and the initial second order moment of the orientation distribution $\mathbf{a}(\mathbf{x}, t = 0)$;
- Solve the equation governing the evolution of $\mathbf{a}$:

$$\dot{\mathbf{a}} = \nabla\mathbf{v} \cdot \mathbf{a} + \mathbf{a} \cdot (\nabla\mathbf{v})^T - 2\mathbf{A} : \mathbf{D} - 4D_r \left(\mathbf{a} - \frac{\mathbf{I}}{2}\right), \text{ in 2D,} \tag{4.130}$$

by using an exact or approximated closure relation.

- Compute the stress tensor:

$$\boldsymbol{\tau} = \boldsymbol{\tau}^f + \boldsymbol{\tau}^r = 2\eta\mathbf{D} + 2\eta N_p (\mathbf{D} : \mathbf{A}) + 2\tilde{\beta} \left(\mathbf{a} - \frac{\mathbf{I}}{2}\right), \text{ in 2D,} \tag{4.131}$$

again using an appropriate closure relation.

4.3.3.4 From Semi-dilute to Concentrated Regimes

Semi-dilute and semi-concentrated regimes have been widely addressed, most of time through the use of phenomenological approaches. The most common approach consists in considering that rod-rod interactions tends to randomize the orientation distribution. Thus, a second diffusion coefficient is introduced so as to account for rods interactions, however, in the present case, that diffusion coefficient should scale with the flow intensity in order to ensure that in the absence of flow, the microstructure does not evolve artificially due to such a diffusion term. In general, the interaction diffusion coefficient D_I is assumed in the general form

$$D_I = C_I \, f(D^{eq}), \tag{4.132}$$

where D^{eq} is related to the second invariant of the rate of strain tensor, i.e., $D^{eq} = \sqrt{\mathbf{D} : \mathbf{D}}$, and C_I is the so-called interaction coefficient. The Folgar & Tucker model considers the simplest dependence $f(D^{eq}) = D^{eq}$.

Obviously, there are finer approaches based on the direct simulation in which the rod-rod interactions are taken explicitly into account. In the concentrated regime, the subjacent physics are richer, and it often involves aggregation. The interested reader can refer to Chap. 2 in [5] and the references therein.

4.3.4 Rigid Clusters Composed of Rigid Rods

The kinematics of rigid and deformable clusters composed of rods were widely described in [5]. When focusing on rigid clusters, it was assumed that hydrodynamic forces (previously introduced) act on the N beads of the rods involved in the cluster. The location of each bead $\mathscr{B}_i$ with respect to the cluster center of gravity $\mathbf{G}$ is given by $L_i\mathbf{p}_i$, where $\mathbf{p}_i$ is the unit vector pointing from $\mathbf{G}$ to $\mathscr{B}_i$, as shown in Fig. 4.3.

A balance of torques yields the cluster rotary velocity. By defining the cluster conformation tensor $\mathbf{c}$ as follows,

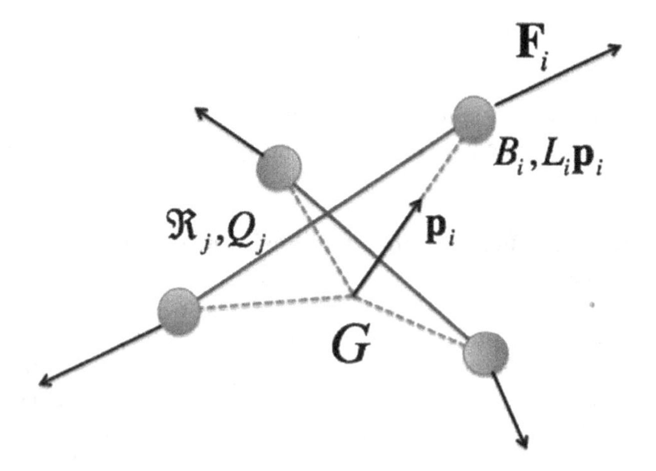

Fig. 4.3 Rigid cluster composed of rods

$$\mathbf{c} = \frac{\sum_{i=1}^{N} L_i^2 \, (\mathbf{p}_i \otimes \mathbf{p}_i)}{\sum_{i=1}^{N} L_i^2}, \tag{4.133}$$

the cluster rotary velocity $\boldsymbol{\omega}$ is given by

$$\boldsymbol{\omega} = (\mathbf{I} - \mathbf{c})^{-1}(\boldsymbol{\varepsilon} : (\nabla \mathbf{v} \cdot \mathbf{c})), \tag{4.134}$$

where $\boldsymbol{\varepsilon}$ is the Levi-Civita permutation tensor. From the cluster rotary velocity $\boldsymbol{\omega}$ given by Eq. (4.134), the rate of change of the conformation tensor $\dot{\mathbf{c}}$ can be calculated [5, 15].

Thus, different clusters having the same conformation tensor $\mathbf{c}$ will have the same rotary velocity. A particular case consists of the tri-dumbell whose rotary velocity calculated from the previous equations coincides with the one associated with the Jeffery solution for the corresponding ellipsoid. In Sect. 4.3.4.1, we describe the rationale in the case of a rigid bi-dumbbell.

In the limit case when the number of rods is large enough, the discrete description given by Eq. (4.133) can be substituted with a continuous one that considers the pdf $\psi(\mathbf{p})$.

A dilute suspension composed of rigid clusters could be described within the kinetic theory framework by using a second pdf, the one related to the clusters distributed in the suspending medium, $\Psi(\mathbf{x}, t, \mathbf{c})$, given the fraction of cluster that, at position $\mathbf{x}$ and time t, has a conformation $\mathbf{c}$. The conformation tensor $\mathbf{c}$ being symmetric and with unit trace, it has, in the general 3D case, 5 independent components, that is, 5 conformational coordinates. The associated Fokker–Planck equation reads as

$$\frac{\partial \Psi}{\partial t} + \nabla_x \cdot (\mathbf{v}\Psi) + \nabla_c \cdot (\dot{\mathbf{c}}\Psi) = 0, \tag{4.135}$$

whose high dimensionality compromises the use of standard mesh-based discretization techniques for solving it, as discussed in [16].

4.3.4.1 Orthogonal Rigid Bi-Dumbbell: Revisiting General Jeffery's Equation

In what follows, we first prove that the kinematics of a rigid ellipse immersed in a flow can be obtained from the analysis of a rigid bi-dumbbell. The Jeffery equation for an ellipse of aspect ratio r (major to minor axes ratio), when considering the decomposition of the velocity gradient in its symmetric (the rate of strain tensor $\mathbf{D}$) and skew-symmetric (the vorticity tensor $\boldsymbol{\Omega}$) components, $\nabla \mathbf{v} = \mathbf{D} + \boldsymbol{\Omega}$ reads as

$$\dot{\mathbf{p}} = \boldsymbol{\Omega} \cdot \mathbf{p} + \mathscr{F}(\mathbf{D} \cdot \mathbf{p} - (\mathbf{p}^T \cdot \mathbf{D} \cdot \mathbf{p})\mathbf{p}), \tag{4.136}$$

with the shape factor $\mathscr{F} = \frac{r^2-1}{r^2+1}$.

In what follows, we consider a rigid system composed of two rods, mutually perpendicular, and having respective lengths of $2L_1$ and $2L_2$ ($L_1 > L_2$), as sketched in Fig. 4.4.

In the present configuration, the hydrodynamic forces acting in bead $L_1\mathbf{p}_1$ and $L_2\mathbf{p}_2$, $\mathbf{F}_1^H$ and $\mathbf{F}_2^H$ respectively read as

$$\mathbf{F}_1^H = \xi L_1(\nabla \mathbf{v} \cdot \mathbf{p}_1 - \dot{\mathbf{p}}_1), \tag{4.137}$$

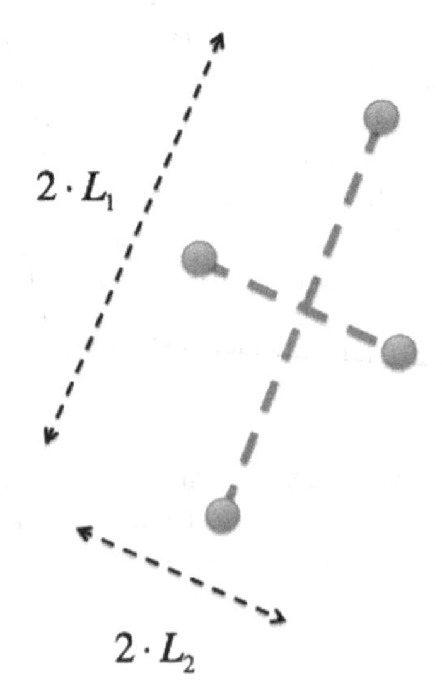

Fig. 4.4 Two-rod rigid cluster representing an elliptic 2D particle

and

$$\mathbf{F}_2^H = \xi L_2(\nabla \mathbf{v} \cdot \mathbf{p}_2 - \dot{\mathbf{p}}_2), \tag{4.138}$$

with $\mathbf{p}_1 \perp \mathbf{p}_2$, and with their rates of change expressed from

$$\begin{cases} \dot{\mathbf{p}}_1 = \boldsymbol{\omega} \times \mathbf{p}_1 \\ \dot{\mathbf{p}}_2 = \boldsymbol{\omega} \times \mathbf{p}_2 \end{cases}, \tag{4.139}$$

where $\boldsymbol{\omega}$ denotes the angular velocity.

The angular momentum balance implies

$$L_1^2 \mathbf{p}_1 \times (\nabla \mathbf{v} \cdot \mathbf{p}_1 - \dot{\mathbf{p}}_1) + L_2^2 \mathbf{p}_2 \times (\nabla \mathbf{v} \cdot \mathbf{p}_2 - \dot{\mathbf{p}}_2) = \mathbf{0}. \tag{4.140}$$

Introducing (4.139) into (4.140) and taking into account that $\mathbf{p}_1 \times \boldsymbol{\omega} \times \mathbf{p}_1 = \boldsymbol{\omega}$ and $\mathbf{p}_2 \times \boldsymbol{\omega} \times \mathbf{p}_2 = \boldsymbol{\omega}$, it results that

$$\boldsymbol{\omega} = \frac{L_1^2}{L_1^2 + L_2^2} (\mathbf{p}_1 \times (\nabla \mathbf{v} \cdot \mathbf{p}_1)) + \frac{L_2^2}{L_1^2 + L_2^2} (\mathbf{p}_2 \times (\nabla \mathbf{v} \cdot \mathbf{p}_2)). \tag{4.141}$$

Thus, $\dot{\mathbf{p}}_1$ can be expressed from

$$\begin{aligned} \dot{\mathbf{p}}_1 &= \boldsymbol{\omega} \times \mathbf{p}_1 \\ &= \frac{L_1^2}{L_1^2 + L_2^2} ((\mathbf{p}_1 \times (\nabla \mathbf{v} \cdot \mathbf{p}_1)) \times \mathbf{p}_1) + \frac{L_2^2}{L_1^2 + L_2^2} ((\mathbf{p}_2 \times (\nabla \mathbf{v} \cdot \mathbf{p}_2) \times \mathbf{p}_1)). \end{aligned} \tag{4.142}$$

Now, applying the triple vector product formula $(\mathbf{a} \times \mathbf{b}) \times \mathbf{c} = -\mathbf{a} \cdot (\mathbf{b} \cdot \mathbf{c}) + \mathbf{b} \cdot (\mathbf{a} \cdot \mathbf{c})$, Eq. (4.142) reads as

$$\begin{aligned} \dot{\mathbf{p}}_1 = & \frac{L_1^2}{L_1^2 + L_2^2} \left(\nabla \mathbf{v} \cdot \mathbf{p}_1 - \left(\mathbf{p}_1^T \cdot \nabla \mathbf{v} \cdot \mathbf{p}_1 \right) \mathbf{p}_1 \right) \\ & - \frac{L_2^2}{L_1^2 + L_2^2} \left(\left(\mathbf{p}_1^T \cdot \nabla \mathbf{v} \cdot \mathbf{p}_2 \right) \mathbf{p}_2 \right). \end{aligned} \tag{4.143}$$

We now develop the last term in Eq. (4.143) to obtain an equation that only contains $\mathbf{p}_1$ in order to compare it with Jeffery's equation. First, we apply the decomposition $\nabla \mathbf{v} = \mathbf{D} + \boldsymbol{\Omega}$ from which the last term in Eq. (4.143) reads as

$$\left(\mathbf{p}_1^T \cdot \nabla \mathbf{v} \cdot \mathbf{p}_2 \right) \mathbf{p}_2 = \left(\mathbf{p}_1^T \cdot \mathbf{D} \cdot \mathbf{p}_2 \right) \mathbf{p}_2 + \left(\mathbf{p}_1^T \cdot \boldsymbol{\Omega} \cdot \mathbf{p}_2 \right) \mathbf{p}_2, \tag{4.144}$$

where the first term in the right hand member, using the fact that $\mathbf{D}$ is symmetric, reads as

$$\left(\mathbf{p}_1^T \cdot \mathbf{D} \cdot \mathbf{p}_2\right) \mathbf{p}_2$$

$$= \mathbf{p}_2 \left(\mathbf{p}_1^T \cdot \mathbf{D} \cdot \mathbf{p}_2\right) = \mathbf{p}_2 \left(\mathbf{p}_2^T \cdot \mathbf{D} \cdot \mathbf{p}_1\right) = (\mathbf{p}_2 \otimes \mathbf{p}_2) \cdot \mathbf{D} \cdot \mathbf{p}_1. \tag{4.145}$$

Now, using the orthogonality of $\mathbf{p}_1$ and $\mathbf{p}_2$, it results that

$$(\mathbf{p}_2 \otimes \mathbf{p}_2) + (\mathbf{p}_1 \otimes \mathbf{p}_1) = \mathbf{I}, \tag{4.146}$$

and thus Eq. (4.145) can be written as

$$\left(\mathbf{p}_1^T \cdot \mathbf{D} \cdot \mathbf{p}_2\right) \mathbf{p}_2 = (\mathbf{I} - \mathbf{p}_1 \otimes \mathbf{p}_1) \cdot \mathbf{D} \cdot \mathbf{p}_1. \tag{4.147}$$

Finally, we consider the second term on the right-hand side of Eq. (4.144), using the fact that $\boldsymbol{\Omega}$ is skew-symmetric

$$\left(\mathbf{p}_1^T \cdot \boldsymbol{\Omega} \cdot \mathbf{p}_2\right) \mathbf{p}_2$$

$$= \mathbf{p}_2 \left(\mathbf{p}_1^T \cdot \boldsymbol{\Omega} \cdot \mathbf{p}_2\right) = -\mathbf{p}_2 \left(\mathbf{p}_2^T \cdot \boldsymbol{\Omega} \cdot \mathbf{p}_1\right) = -(\mathbf{p}_2 \otimes \mathbf{p}_2) \cdot \boldsymbol{\Omega} \cdot \mathbf{p}_1, \tag{4.148}$$

that is,

$$\left(\mathbf{p}_1^T \cdot \boldsymbol{\Omega} \cdot \mathbf{p}_2\right) \mathbf{p}_2 = -(\mathbf{I} - \mathbf{p}_1 \otimes \mathbf{p}_1) \cdot \boldsymbol{\Omega} \cdot \mathbf{p}_1. \tag{4.149}$$

Now, coming back to Eq. (4.143), we obtain

$$\dot{\mathbf{p}}_1 = \boldsymbol{\Omega} \cdot \mathbf{p}_1 + \frac{L_1^2 - L_2^2}{L_1^2 + L_2^2} \mathbf{D} \cdot \mathbf{p}_1 - \frac{L_1^2 - L_2^2}{L_1^2 + L_2^2} \left(\mathbf{p}_1^T \cdot \mathbf{D} \cdot \mathbf{p}_1\right) \mathbf{p}_1, \tag{4.150}$$

which corresponds exactly to the Jeffery expression for an ellipse of aspect ratio $\frac{L_1}{L_2}$.

4.3.5 *Extensible Rods*

We consider the kinematics of an extensible rod, whose reference length is $2L^0$. In first gradient flows (at the rod scale), only the rod's extension and its rotation are activated by the flow kinematics. At time t, the rod, aligned in the direction $\mathbf{p}$, is represented by an elastic spring of length $2L$ and rigidity $\mathscr{K}$ equipped with two beads at its extremities where the hydrodynamic forces act. As previously considered, the hydrodynamic force scales with the fluid/bead relative velocity, the former given by $\mathbf{v}_0 + \nabla \mathbf{v} \cdot \mathbf{p}L$ and the latter by $\mathbf{v}_G + \dot{\mathbf{p}}L + \mathbf{p}\dot{L}$, where $\mathbf{v}_0$ is the unperturbed fluid velocity at the rod's center of gravity and $\mathbf{v}_G$ that center of gravity's own velocity. A sketch of the elastic rod and the forces applied to it is depicted in Fig. 4.5.

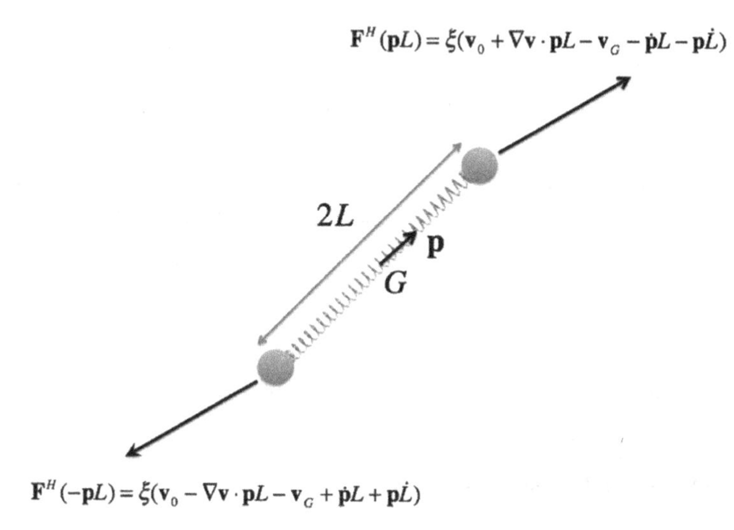

Fig. 4.5 Extensible rod immersed in a flow

The system is assumed inertialess, implying that the equilibrium of forces and torques applies. The first reads as $\mathbf{F}^H(\mathbf{p}L) + \mathbf{F}^H(-\mathbf{p}L) = \mathbf{0}$, which implies that $\mathbf{v}_0 = \mathbf{v}_G$, that is, the rod's centre of gravity moves with the fluid.

Now, to prevent a resultant torque, force $\mathbf{F}^H(\mathbf{p}L)$ must align with $\mathbf{p}$, i.e., $\mathbf{F}^H(\mathbf{p}L) = \lambda\mathbf{p}$, $\lambda \in \mathbb{R}$. Thus, it results that

$$\mathbf{F}^H(\mathbf{p}L) = \xi(\nabla\mathbf{v}\cdot\mathbf{p}L - \dot{\mathbf{p}}L - \mathbf{p}\dot{L}) = \lambda\mathbf{p}, \tag{4.151}$$

which, multiplying by $\mathbf{p}$ and taking into account that $\mathbf{p}\cdot\mathbf{p} = 1$, and consequently $\mathbf{p}\cdot\dot{\mathbf{p}} = 0$, yields

$$\xi(\nabla\mathbf{v} : (\mathbf{p}\otimes\mathbf{p})L - \dot{L}) = \lambda. \tag{4.152}$$

Introducing this expression into Eq. (4.151), it reads as

$$\xi(\nabla\mathbf{v}\cdot\mathbf{p}L - \dot{\mathbf{p}}L - \mathbf{p}\dot{L}) = \xi(\nabla\mathbf{v} : (\mathbf{p}\otimes\mathbf{p})L - \dot{L})\mathbf{p}, \tag{4.153}$$

and grouping terms

$$\dot{\mathbf{p}} = \nabla\mathbf{v}\cdot\mathbf{p} - \nabla\mathbf{v} : (\mathbf{p}\otimes\mathbf{p})\mathbf{p}, \tag{4.154}$$

which is nothing other than the standard Jeffery expression for ellipsoids of infinite aspect ratio (rods) widely discussed in the previous sections.

Now, by equating the force acting on the beads, λ, with the one installed in the spring, it results that

$$2\mathscr{K}(L - L^0) = \xi(\nabla\mathbf{v} : (\mathbf{p}\otimes\mathbf{p})L - \dot{L}), \tag{4.155}$$

or

$$\dot{L} = -\frac{2\mathscr{K}}{\xi}(L - L^0) + \nabla\mathbf{v} : (\mathbf{p} \otimes \mathbf{p})L. \tag{4.156}$$

Thus, the kinematics of an elastic dumbbell of reference length $2L^0$ immersed in a first gradient flow and described by its orientation $\mathbf{p}$ and length $2L$ are given by

$$\begin{cases} \dot{\mathbf{p}} = \nabla\mathbf{v} \cdot \mathbf{p} - \nabla\mathbf{v} : (\mathbf{p} \otimes \mathbf{p})\mathbf{p} \\ \dot{L} = -\frac{2\mathscr{K}}{\xi}(L - L^0) + \nabla\mathbf{v} : (\mathbf{p} \otimes \mathbf{p})L \end{cases}. \tag{4.157}$$

Before closing this section, we address the case of a deformable bi-dumbbell.

4.3.5.1 Orthogonal Elastic Bi-Dumbell

In the present configuration, and considering that as proven earlier, the centre of gravity moves with the fluid, the hydrodynamic forces acting in bead $L_1\mathbf{p}_1$ and $L_2\mathbf{p}_2$, $\mathbf{F}_1^H$ and $\mathbf{F}_2^H$, respectively read as

$$\mathbf{F}_1^H = \xi(\nabla\mathbf{v} \cdot \mathbf{p}_1 L_1 - \dot{\mathbf{p}}_1 L_1 - \mathbf{p}_1 \dot{L}_1), \tag{4.158}$$

and

$$\mathbf{F}_2^H = \xi(\nabla\mathbf{v} \cdot \mathbf{p}_2 L_2 - \dot{\mathbf{p}}_2 L_2 - \mathbf{p}_2 \dot{L}_2), \tag{4.159}$$

Again, with $\mathbf{p}_1 \perp \mathbf{p}_2$, and with their rates of change expressed by

$$\begin{cases} \dot{\mathbf{p}}_1 = \boldsymbol{\omega} \times \mathbf{p}_1 \\ \dot{\mathbf{p}}_2 = \boldsymbol{\omega} \times \mathbf{p}_2 \end{cases}. \tag{4.160}$$

The angular momentum balance now implies that

$$L_1^2\mathbf{p}_1 \times (\nabla\mathbf{v} \cdot \mathbf{p}_1 - \dot{\mathbf{p}}_1) + L_2^2\mathbf{p}_2 \times (\nabla\mathbf{v} \cdot \mathbf{p}_2 - \dot{\mathbf{p}}_2) = \mathbf{0}, \tag{4.161}$$

which coincides with the expression obtained in the case of orthogonal rigid bi-dumbbells, proving the validity of the Jeffery equation in the case of an orthogonal elastic bi-dumbell.

However, in the present case, we need extra-equations giving the spring extensions $\dot{L}_1$ and $\dot{L}_2$. For that purpose, we consider the Jeffery equation (4.136) with $\mathscr{F} = \frac{L_1^2 - L_2^2}{L_1^2 + L_2^2}$, i.e.,

$$\dot{\mathbf{p}}_1 = \boldsymbol{\Omega} \cdot \mathbf{p}_1 + \mathscr{F}\left(\mathbf{D} \cdot \mathbf{p}_1 - \left(\mathbf{p}_1^T \cdot \mathbf{D} \cdot \mathbf{p}_1\right)\mathbf{p}_1\right), \tag{4.162}$$

and introduce it into the expression of the hydrodynamic force acting on bead $\mathbf{p}_1 L_1$ (4.158)

$$\mathbf{F}_1^H = \xi(\nabla \mathbf{v} \cdot \mathbf{p}_1 L_1 - \dot{\mathbf{p}}_1 L_1 - \mathbf{p}_1 \dot{L}_1)$$

$$= \xi \left(\nabla \mathbf{v} \cdot \mathbf{p}_1 L_1 - \boldsymbol{\Omega} \cdot \mathbf{p}_1 L_1 - \mathscr{F} \left(\mathbf{D} \cdot \mathbf{p}_1 L_1 - \left(\mathbf{p}_1^T \cdot \mathbf{D} \cdot \mathbf{p}_1\right) \mathbf{p}_1 L_1\right) - \mathbf{p}_1 \dot{L}_1\right)$$

$$= \xi L_1 \left((1 - \mathscr{F})\mathbf{D} \cdot \mathbf{p}_1 + \mathscr{F} \left(\mathbf{p}_1^T \cdot \mathbf{D} \cdot \mathbf{p}_1\right) \mathbf{p}_1\right) - \xi \mathbf{p}_1 \dot{L}_1, \tag{4.163}$$

which proves that the force acting on bead $\mathbf{p}_1 L_1$ aligns in direction $\mathbf{p}_1$ as soon as $\mathscr{F} = 1$.

The projection of force $\mathbf{F}_1^H$ in the direction $\mathbf{p}_1$ is the one that causes the spring extension, i.e.,

$$2\mathscr{K}_1(L_1 - L_1^0) = \xi L_1 \left((1 - \mathscr{F})\mathbf{p}_1^T \cdot \mathbf{D} \cdot \mathbf{p}_1 + \mathscr{F} \left(\mathbf{p}_1^T \cdot \mathbf{D} \cdot \mathbf{p}_1\right)\right) - \xi \dot{L}_1 =$$

$$\xi L_1 \mathbf{p}_1^T \cdot \mathbf{D} \cdot \mathbf{p}_1 - \xi \dot{L}_1, \tag{4.164}$$

that is,

$$\dot{L}_1 = -\mathscr{K}_1^*(L_1 - L_1^0) + L_1 \mathbf{p}_1^T \cdot \mathbf{D} \cdot \mathbf{p}_1, \tag{4.165}$$

with $\mathscr{K}_1^* = \frac{2\mathscr{K}_1}{\xi}$.

Obviously repeating the rationale for $\mathbf{F}_2^H$, it results that

$$\dot{L}_2 = -\mathscr{K}_2^*(L_2 - L_2^0) + L_2 \mathbf{p}_2^T \cdot \mathbf{D} \cdot \mathbf{p}_2, \tag{4.166}$$

Again, with $\mathscr{K}_2^* = \frac{2\mathscr{K}_2}{\xi}$.

4.3.6 Dilute Polymers Solutions

By assuming extensible bead connectors instead of rigid dumbbells, the resulting models accurately describe some viscoelastic features of dilute polymer solutions. In this case, linear macromolecules are crudely represented by two beads connected by an extensible (linear or non-linear) spring.

In that which follows, we summarize some of the most frequent microstructural models based on this representation. We then start by considering models based on a single connector, to finish with consideration of the more general (and more complex) Multi-Bead-Spring – MBS.

4.3.6.1 Linear and Nonlinear Elastic Dumbells

We start by considering two beads connected by an elastic spring (Fig. 4.5) of null reference length, i.e., $L^0 = 0$. In fact, the spring reference length represents the equilibrium end-to-end distance of a macromolecule idealized as a sequence of beads distributed along its contour and connected by rigid rods. Relative rotation of consecutive rods is allowed at the common bead. In the absence of forces, the end-to-end averaged distance vanishes, justifying the fact that we have considered a null reference length of the elastic spring. For a complete description, the interested reader can refer to [17].

Linear Spring

Considering again the system depicted in Fig. 4.5, neglecting Brownian forces, assuming a linear spring with rigidity $\mathscr{K}$ and null reference length, and proceeding in a similar way as in the previous section, we obtain model (4.157) with $L^0 = 0$

$$\begin{cases} \dot{\mathbf{p}} = \nabla \mathbf{v} \cdot \mathbf{p} - \nabla \mathbf{v} : (\mathbf{p} \otimes \mathbf{p})\mathbf{p} \\ \dot{L} = -\frac{2\mathscr{K}}{\xi} L + \nabla \mathbf{v} : (\mathbf{p} \otimes \mathbf{p})L) \end{cases} . \tag{4.167}$$

The time evolution of the connector $\mathbf{q} = \mathbf{p}L$, taking into account both expressions in (4.167),

$$\dot{\mathbf{q}} = \dot{\mathbf{p}}L + \mathbf{p}\dot{L} = \nabla \mathbf{v} \cdot \mathbf{q} - \frac{2\mathscr{K}}{\xi}\mathbf{q}, \tag{4.168}$$

in which we identify the affine deformation $\nabla \mathbf{v} \cdot \mathbf{q}$, which is now no longer constrained by a constant length, as was the case when considering the unit vector $\mathbf{p}$ and the elastic term proportional to the spring stretching $\mathbf{q}$ (remember that the spring reference length is zero).

Now, the mesoscopic model involving the pdf $\Psi(\mathbf{x}, t, \mathbf{q})$ results in the Fokker–Planck equation

$$\frac{\partial \Psi}{\partial t} + \nabla_x \cdot (\mathbf{v}\Psi) + \nabla_q \cdot (\dot{\mathbf{q}}\Psi) = 0, \tag{4.169}$$

where, for the moment, Brownian effects are neglected. The first two terms in the previous equation can be grouped into the material derivative $\frac{d}{dt}$, and thus it allows us to rewrite it as

$$\frac{d\Psi}{dt} + \nabla_q \cdot (\dot{\mathbf{q}}\Psi) = 0. \tag{4.170}$$

To define the meso-to-macro bridge, we define the conformation tensor $\mathbf{c}(\mathbf{x}, t)$ from

$$\mathbf{c}(\mathbf{x}, t) = \int_{\mathbb{R}^3} \mathbf{q} \otimes \mathbf{q}\Psi(\mathbf{x}, t, \mathbf{q})\, d\mathbf{q}. \tag{4.171}$$

In order to determine the equation governing its evolution, $\dot{\mathbf{c}}$, we multiply both members of Eq. (4.170) by $\mathbf{q} \otimes \mathbf{q}$, and then we integrate into the configurational space in which $\mathbf{q}$ is defined, $\mathbb{R}^3$, making use of the integration by parts, to obtain

$$\frac{d\mathbf{c}}{dt} = \nabla\mathbf{v} \cdot \mathbf{c} + \mathbf{c} \cdot (\nabla\mathbf{v})^T - \frac{4\mathscr{K}}{\xi}\mathbf{c}. \tag{4.172}$$

To consider Brownian effects, we write the equilibrium of forces applied to the two beads, the first located at $\mathbf{r}_1$ and the second at $\mathbf{r}_2$, with the connector $\mathbf{q}$ defined by $\mathbf{q} = \mathbf{r}_2 - \mathbf{r}_1$:

$$-\xi(\dot{\mathbf{r}}_2 - \mathbf{v}_0 - \nabla\mathbf{v} \cdot \mathbf{r}_2) - K_b T \frac{\partial}{\partial \mathbf{r}_2} \ln(\Psi) - \mathbf{F}^C = \mathbf{0}, \tag{4.173}$$

where ξ is the drag coefficient, $\mathbf{v}_0$ is the fluid velocity at the origin of coordinates, K_b is the Boltzmann constant, T is the absolute temperature and $\mathbf{F}^C$ is the elastic force in the spring. Here, because of the assumed linearity, $\mathbf{F}^C$ is proportional to its length and acts along the direction of the connector (spring), i.e., $\mathbf{F}^C = \mathscr{K}\mathbf{q}$. We consider the random force involving the logarithm of the the distribution function Ψ to recover a diffusion term in the associated Fokker–Planck equation.

The balance of forces at the other bead reads as

$$-\xi(\dot{\mathbf{r}}_1 - \mathbf{v}_0 - \nabla\mathbf{v} \cdot \mathbf{r}_1) - K_b T \frac{\partial}{\partial \mathbf{r}_1} \ln(\Psi) + \mathbf{F}^C = \mathbf{0}. \tag{4.174}$$

By subtracting the second one from the first one, taking into account

$$\begin{cases} \frac{\partial}{\partial \mathbf{r}_2} = \frac{\partial}{\partial \mathbf{q}} \frac{\partial \mathbf{q}}{\partial \mathbf{r}_2} = \frac{\partial}{\partial \mathbf{q}} \\ \frac{\partial}{\partial \mathbf{r}_1} = \frac{\partial}{\partial \mathbf{q}} \frac{\partial \mathbf{q}}{\partial \mathbf{r}_1} = -\frac{\partial}{\partial \mathbf{q}} \end{cases}, \tag{4.175}$$

it results that

$$-\xi(\dot{\mathbf{q}} - \nabla\mathbf{v} \cdot \mathbf{q}) - 2K_b T \frac{\partial}{\partial \mathbf{q}} \ln(\Psi) - 2\mathbf{F}^C = \mathbf{0}, \tag{4.176}$$

from which the expression of $\dot{\mathbf{q}}$ can be extracted

$$\dot{\mathbf{q}} = \nabla\mathbf{v} \cdot \mathbf{q} - \frac{2K_b T}{\xi} \frac{\partial}{\partial \mathbf{q}} \ln(\Psi) - \frac{2}{\xi}\mathbf{F}^C, \tag{4.177}$$

and using the linear behavior of the elastic spring,

$$\dot{\mathbf{q}} = \nabla\mathbf{v} \cdot \mathbf{q} - \frac{2K_b T}{\xi} \frac{\partial}{\partial \mathbf{q}} \ln(\Psi) - \frac{2\mathscr{K}}{\xi}\mathbf{q}, \tag{4.178}$$

It can be noticed that, in the absence of Brownian effects and for a linear spring, Eq. (4.178) reduces to the one previously obtained, Eq. (4.168).

The mesoscopic model is defined, as before, by

$$\frac{\partial \Psi}{\partial t} + \nabla_x \cdot (\mathbf{v}\Psi) + \nabla_q \cdot (\dot{\mathbf{q}}\Psi) = 0, \tag{4.179}$$

which, making use of the material derivative, reduces to

$$\frac{d\Psi}{dt} + \nabla_q \cdot (\dot{\mathbf{q}}\Psi) = 0, \tag{4.180}$$

where $\dot{\mathbf{q}}$ is now given by Eq. (4.178), which includes Brownian effects.

In order to determine the equation governing the evolution of the conformation tensor $\dot{\mathbf{c}}$, when Brownian effects are retained in the model, we multiply both members of Eq. (4.180) by $\mathbf{q} \otimes \mathbf{q}$, and then we integrate into the configurational space $\mathbb{R}^3$, making use of the integration by parts, to obtain

$$\frac{d\mathbf{c}}{dt} = \nabla\mathbf{v} \cdot \mathbf{c} + \mathbf{c} \cdot (\nabla\mathbf{v})^T - \frac{4\mathcal{K}}{\xi}\mathbf{c} + \frac{4K_bT}{\xi}\mathbf{I}. \tag{4.181}$$

If the extra-stress is assumed given by Kramer's expression, with n the number of elastic dumbbells per unit volume, it results that

$$\boldsymbol{\tau} + nK_bT\mathbf{I} = n\langle \mathbf{F}^c \otimes \mathbf{q} \rangle = n\mathcal{K}\mathbf{c}, \tag{4.182}$$

where the term $nK_bT\mathbf{I}$ is isotropic, and thus it does not have any rheological consequence.

By substituting expression (4.182) into Eq. (4.181), it results that

$$\frac{1}{n\mathcal{K}}\left(\frac{d\boldsymbol{\tau}}{dt} - \nabla\mathbf{v} \cdot \boldsymbol{\tau} - \boldsymbol{\tau} \cdot (\nabla\mathbf{v})^T\right) + \frac{4}{n\xi}\boldsymbol{\tau} = 2nK_bT\mathbf{D}, \tag{4.183}$$

where the upper-convected derivative $\frac{\delta}{\delta t}$ can be identified, leading to

$$\frac{1}{n\mathcal{K}}\frac{\delta\boldsymbol{\tau}}{\delta t} + \frac{4}{n\xi}\boldsymbol{\tau} = 2nK_bT\mathbf{D}, \tag{4.184}$$

which, by introducing the parameters

$$\begin{cases} \lambda = \frac{\xi}{4\mathcal{K}} \\ \eta_p = \frac{n^2K_bT\xi}{4} \end{cases}, \tag{4.185}$$

leads to the viscoelastic Oldroyd-B model

$$\lambda\frac{\delta\boldsymbol{\tau}}{\delta t} + \boldsymbol{\tau} = 2\eta_p\mathbf{D}. \tag{4.186}$$

Nonlinear Spring

When considering a nonlinear spring's behavior, the time evolution of the connector $\mathbf{q}$ reads as

$$\dot{\mathbf{q}} = \nabla \mathbf{v} \cdot \mathbf{q} - \frac{2\mathscr{K}(q)}{\xi}\mathbf{q}, \tag{4.187}$$

with $q = \|\mathbf{q}\|$.

Now, the mesoscopic model involving the pdf $\Psi(\mathbf{x}, t, \mathbf{q})$ results in the Fokker–Planck equation

$$\frac{\partial \Psi}{\partial t} + \nabla_x \cdot (\mathbf{v}\Psi) + \nabla_q \cdot (\dot{\mathbf{q}}\Psi) = 0, \tag{4.188}$$

or, using the material derivative $\frac{d}{dt}$,

$$\frac{d\Psi}{dt} + \nabla_q \cdot (\dot{\mathbf{q}}\Psi) = 0. \tag{4.189}$$

The non linear term $\mathscr{K}(q)$ makes it impossible, in the general case, to obtain a closed equation for the evolution of the conformation tensor, and thus for the extra-stress tensor (constitutive equation). The introduction of particular closures makes possible the obtention of closed equations, as is the case with FENE-P models (see Chap. 1 in [5]).

Thus, the microscopic counterpart of the viscoelastic Oldroyd-B model can be derived by using a linear spring, and the so-called FENE (Finitely Extensible Non Linear Elastic) model is obtained by considering the connector force

$$\mathbf{F}^c = \mathscr{K}(q)\mathbf{q} = \frac{1}{1 - \frac{\|\mathbf{q}\|^2}{b}}\mathbf{q}, \tag{4.190}$$

where $\sqrt{b}$ is the maximum stretching of the spring connector.

4.3.6.2 Multi-Bead-Spring Models

As depicted in Fig. 4.6, the Multi-Bead-Spring – MBS – chain consists of $N + 1$ beads connected by N springs. The bead serves as an interaction point with the solvent, and the spring contains the local stiffness information depending on local stretching (see [18, 19] and the references therein for additional details).

The dynamics of the chain is governed by hydrodynamic, Brownian and connector (spring) forces. If we denote by $\dot{\mathbf{r}}_j$ the velocity of the bead j and by $\dot{\mathbf{q}}_j$ the velocity of the spring connector j, then we obtain

$$\dot{\mathbf{q}}_j = \dot{\mathbf{r}}_{j+1} - \dot{\mathbf{r}}_j; \quad j = 1, \ldots, N. \tag{4.191}$$

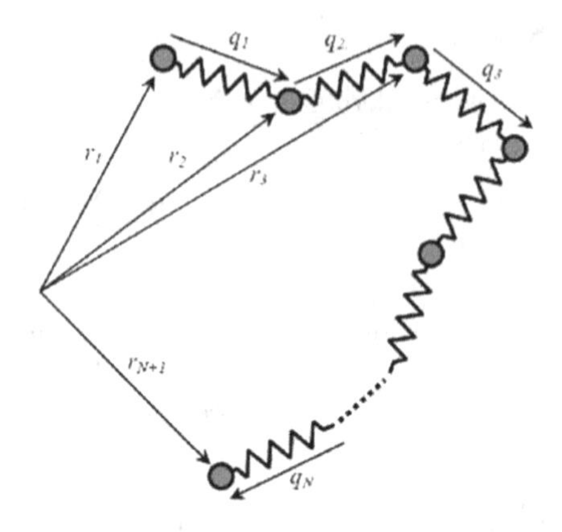

Fig. 4.6 MBS model of a polymer chain

The kinetic theory approach consists in introducing the microstructural conformation from the pdf $\Psi(\mathbf{x}, t, \mathbf{r}_1, \ldots, \mathbf{r}_{N+1})$, or in an equivalent manner, making use of the connector vectors, $\Psi(\mathbf{x}, t, \mathbf{q}_1, \ldots, \mathbf{q}_N)$.

The force balance at each bead reads as

$$\underbrace{-\xi(\dot{\mathbf{r}}_j - \mathbf{v}_0 - \nabla\mathbf{v}\cdot\mathbf{r}_j)}_{\text{Hydrodynamic contribution}} \underbrace{- K_b T \frac{\partial}{\partial \mathbf{r}_j}\ln(\Psi)}_{\text{Brownian contribution}} \underbrace{+ \mathbf{F}^c_j - \mathbf{F}^c_{j-1}}_{\text{Connector forces}} = \mathbf{0}, \tag{4.192}$$

where ξ is the drag coefficient, $\mathbf{v}$ is the velocity field, $\mathbf{v}_0$ is the fluid velocity and the origin of coordinates, K_b is the Boltzmann constant and T is the absolute temperature.

From Eqs. (4.191) and (4.192), it results that

$$\dot{\mathbf{q}}_j = \nabla\mathbf{v}\cdot\mathbf{q}_j - \frac{1}{\xi}\sum_{k=1}^{N}\mathbf{A}_{jk}\left(K_b T \nabla_{\mathbf{q}_k}\ln(\Psi) + \mathbf{F}^c_k\right), \tag{4.193}$$

where $\nabla_{\mathbf{q}_k} = \frac{\partial}{\partial \mathbf{q}_k}$ and $\mathbf{A}$ is the Rouse matrix with components

$$\mathbf{A}_{jk} = \begin{cases} 2 & \text{if } k = j \\ -1 & \text{if } k = j \pm 1 \\ 0 & \text{otherwise} \end{cases}. \tag{4.194}$$

In the Rouse model, the connector force $\mathbf{F}^c$ is a linear function of the connector vector, but FENE springs previously introduced can be also used.

The pdf involving the physical coordinates (space and time) and the conformational coordinates (connector vectors) $\Psi(\mathbf{x}, t, \mathbf{q}_1, \ldots, \mathbf{q}_N)$ is defined in a space of dimension $3 + 1 + 3N$, implying the so-called curse of dimensionality when N increases.

The conservation balance related to the distribution function $\Psi(\mathbf{x}, t, \mathbf{q}_1, \ldots, \mathbf{q}_N)$ reads as

$$\frac{\partial \Psi}{\partial t} + \nabla_x \cdot (\mathbf{v}\Psi) = -\sum_{j=1}^{N} \nabla_{\mathbf{q}_j} \cdot (\dot{\mathbf{q}}_j \ \Psi). \tag{4.195}$$

By introducing Eq. (4.193), it results that

$$\begin{aligned} \tfrac{\partial \Psi}{\partial t} + \nabla_x \cdot (\mathbf{v}\Psi) = -\sum_{j=1}^{N} \nabla_{\mathbf{q}_j} \cdot \left(\left(\nabla\mathbf{v} \cdot \mathbf{q}_j - \frac{1}{\xi}\sum_{k=1}^{N} \mathbf{A}_{jk}\ \mathbf{F}_k^c\right)\Psi\right) + \\ + \tfrac{K_b T}{\zeta} \sum_{j=1}^{N} \nabla_{\mathbf{q}_j} \cdot \left(\sum_{k=1}^{N} \mathbf{A}_{jk} \nabla_{\mathbf{q}_k} \Psi\right). \end{aligned} \tag{4.196}$$

This kind of model seems particularly suitable for representing molecular chains composed of rigid segments, as in the case of DNA. Other variants can be defined, for example, through restricting the free rotation of each two consecutive rods by introducing a rotational spring at each bead. In all of these models, the expression of the rods' rotational velocity (4.193) must be reevaluated depending on the considered physics.

4.3.7 *Polymer Melts*

In most polymer processing operations, such as injection molding, film blowing and extrusion, the polymers are in the molten state. A widely applied class of molecular-based models for concentrated polymer solutions and melts relies on the notion of reptation motion.

The key idea of this model is the application of the reptation mechanism introduced by De-Gennes [20] to a tube (along which the molecule can move) in order to describe the viscoelastic behavior of entangled polymers. The molecule is described as sliding or reptating through a tube whose contours are defined by the locus of entanglements with neighboring molecules. The motion of a molecular chain in any direction other than the one defined by the tube axis is strongly restricted, except at both tube ends, where it can move in any possible direction. The tube moves itself thanks to two mechanisms: (i) by means of the motion of the central chain itself, which partially leaves its original tube, to extend it in other directions, and (ii) by the fluctuation induced by the motions of the neighbor chains defining the tube's lateral border.

In addition to the reptation mechanism, the Doi-Edwards model [21] assumes affine tube deformation induced by the macroscopic flow, but neglects other phenomena, like the stretch of the chain, the Convective Constraint Release – CCR – or the double reptation.

Within the reptation picture and these assumptions, the dynamics of a single tube segment represented by the unit $\mathbf{u}$ (tangent to the tube) are given by the affine deformation

$$\dot{\mathbf{u}} = \nabla \mathbf{v} \cdot \mathbf{u} - (\nabla \mathbf{v} : (\mathbf{u} \otimes \mathbf{u}))\mathbf{u}. \tag{4.197}$$

The molecule defining and occupying the tube at the initial time, is equipped with a contour length coordinate s, where $s = 0$ and $s = 1$ represent the chain ends.

The distribution function involved in the Doi-Edwards model depends on the physical space and time, and also on the conformation coordinates $\mathbf{u}$ and s, $\Psi(\mathbf{x}, t, \mathbf{u}, s)$, where $\Psi(\mathbf{x}, t, \mathbf{u}, s) d\mathbf{u} ds$ represents the joint probability that, at position $\mathbf{x}$ and time t, a tube segment having the orientation in the interval $[\mathbf{u}, \mathbf{u} + d\mathbf{u}]$ contains the chain segment labelled in the interval $[s, s + ds]$. Thus, the configuration space results in $\mathcal{S} \times [0, 1]$.

The balance of $\Psi(\mathbf{x}, \mathbf{u}, s, t)$ results in

$$\frac{\partial \Psi}{\partial t} + \nabla_x \cdot (\mathbf{v}\Psi) + \nabla_u \cdot (\dot{\mathbf{u}}\Psi) + \nabla_s(q_s) = 0, \tag{4.198}$$

where the reptation flux can be modeled from a diffusion term

$$q_s = -D_r \frac{\partial \Psi}{\partial s}, \tag{4.199}$$

whose diffusion coefficient D_r is related to the reptation characteristic time τ_r for a chain to come out of the tube by reptation,

$$D_r = \frac{1}{\pi^2 \tau_r}. \tag{4.200}$$

Thus, Eq. (4.198) can be written as

$$\frac{\partial \Psi}{\partial t} + \nabla_x \cdot (\mathbf{v}\Psi) + \nabla_u \cdot (\dot{\mathbf{u}}\Psi) - D_r \frac{\partial^2 \Psi}{\partial s^2} = 0. \tag{4.201}$$

To solve the Fokker–Planck equation, one needs to prescribe appropriate boundary conditions at the conformational domain boundaries $s = 0$ and $s = 1$ (the orientation coordinate being defined on the unit sphere does not require enforcing any boundary condition). The usual choice is assuming an isotropic orientation distribution at both ends (even if other choices could be more pertinent in some situations), which reads as

$$\Psi(\mathbf{x}, t, \mathbf{u}, s = 0) = \Psi(\mathbf{x}, t, \mathbf{u}, s = 1) = \frac{1}{4\pi}. \tag{4.202}$$

Knowing the distribution function, the stress can be computed from

$$\boldsymbol{\tau}(\mathbf{x}, t) = G \int_{\mathscr{S} \times [0,1]} \mathbf{u} \otimes \mathbf{u}\, \Psi(\mathbf{x}, t, \mathbf{u}, s)\, d\mathbf{u} ds, \tag{4.203}$$

where G is an elastic modulus.

However, the Doi-Edwards model seems to be too simplistic to take into account certain features observed experimentally. More advanced kinetic theory models were proposed, including chain stretching, double reptation and convective constraint release.

4.3.8 Liquid Crystalline Polymers

Important difficulties persist in the numerical calculation of flows of complex fluids consisting of highly anisotropic particles. Most of these difficulties stem from the necessity to capture the rich dynamics fluid's behavior, which spans both isotropic and anisotropic fluid microstructures (nematic, chiral nematic, smectic, etc.). For example, models describing single particle orientation in a dilute suspension, i.e., in the absence of inter-particle interaction, such as the one intensively described earlier based on the Jeffery model, are not able to describe the behavior of a nematic suspension.

The successful treatment of collective flow-orientation phenomena appearing at high concentrations requires the consideration of interparticle interaction, either explicitly, as in microscopic liquid-cristal models – LC – or in a mean-field sense. There are notable examples of mean field approaches to anisotropic particle interaction described within the kinetic theory framework.

Different closure approximations have been proposed for alleviating the numerical simulation of LC models by proceeding at the macroscopic scale. The impact of these approximations on the rheological response can be significant, even if some of them retain the main features of the kinetic theory model.

An appealing choice that allows for circumventing the introduction of such approximations consists in directly solving the Fokker–Planck equation related to the kinetic theory description of LC models. In this section, we consider the Fokker–Planck equation associated with Doi's model.

In order to describe the fluid's microstructure, we first define the unit vector $\mathbf{u}$ describing the orientation of the liquid crystal (LC) molecule axis. Now, the distribution function that gives the probability of finding, at a certain point $\mathbf{x}$ and time t, molecules aligned in the direction $\mathbf{u}$ is represented by $\Psi(\mathbf{x}, \mathbf{u}, t)$, defined by: $\Psi(\mathbf{x}, \mathbf{u}, t) : \mathbb{R}^3 \times \mathscr{S} \times \mathbb{R}^+ \rightarrow \mathbb{R}^+$. The dynamics of the evolution of this distribution function is induced by (i) hydrodynamic effects, (ii) Brownian effects and (iii) LC-molecule interaction described by a nematic potential. The corresponding Fokker–Planck equation related to Doi's model writes

$$\frac{\partial \Psi}{\partial t} + \nabla_x \cdot (\mathbf{v}\Psi)$$

$$= -\nabla_u \cdot (\dot{\mathbf{u}}\,\Psi) + \nabla_u \cdot (D_r \nabla_u \Psi) + \nabla_u \cdot \left(D_r \Psi \nabla_u \left(\frac{V(\mathbf{u})}{K_b T} \right) \right), \tag{4.204}$$

where the last two terms describe diffusion and nematic effects, K_b is Boltzmann's constant, D_r the diffusion coefficient describing the Brownian effects, T the absolute temperature and $\mathbf{v}$ the fluid velocity field. The rotary velocity is dictated by an affine deformation assumption

$$\dot{\mathbf{u}} = \nabla \mathbf{v} \cdot \mathbf{u} - (\nabla \mathbf{v} : (\mathbf{u} \otimes \mathbf{u}))\,\mathbf{u}. \tag{4.205}$$

We can note that removing the last term in Eq. (4.204) involving the nematic potential $V(\mathbf{u})$ reduces to the kinetic theory description of rods suspensions previously described.

A form frequently used for the nematic potential is

$$V(\mathbf{u}) = -\frac{3}{2} U K_b T (\mathbf{u} \otimes \mathbf{u}) : \mathbf{S}, \tag{4.206}$$

where U is a dimensionless interaction potential and the second order traceless orientation tensor $\mathbf{S}$ is defined by

$$\mathbf{S} = \langle \mathbf{u} \otimes \mathbf{u} \rangle - \frac{\mathbf{I}}{3}, \tag{4.207}$$

with the averaging $\langle \bullet \rangle$ defined on the unit sphere $\mathcal{S}$ by the second moment of the orientation distribution (as was the case when considering rigid rods)

$$\langle \mathbf{u} \otimes \mathbf{u} \rangle = \int_{\mathcal{S}} \mathbf{u} \otimes \mathbf{u}\, \Psi(\mathbf{x}, \mathbf{u}, t)\, d\mathbf{u}. \tag{4.208}$$

To illustrate the physical meaning of the nematic potential, we consider a 2D molecular orientation distribution (in the absence of macroscopic flow) where most molecules are fully aligned in a certain direction, for example, in the x-direction. In this case, the different entities appearing in the nematic potential result in

$$\langle \mathbf{u} \otimes \mathbf{u} \rangle = \left\langle \begin{pmatrix} \cos^2\varphi & \cos\varphi \sin\varphi \\ \cos\varphi \sin\varphi & \sin^2\varphi \end{pmatrix} \right\rangle \approx \begin{pmatrix} 1 & 0 \\ 0 & 0 \end{pmatrix}, \tag{4.209}$$

which implies that

$$\mathbf{S} = \langle \mathbf{u} \otimes \mathbf{u} \rangle - \frac{\mathbf{I}}{2} \approx \begin{pmatrix} \frac{1}{2} & 0 \\ 0 & -\frac{1}{2} \end{pmatrix}, \tag{4.210}$$

and thus,

$$V(\mathbf{u}) = -\frac{3}{2} U K_b T (\mathbf{u} \otimes \mathbf{u}) : \mathbf{S} = -\beta \left(\frac{\cos^2(\varphi)}{2} - \frac{\sin^2(\varphi)}{2} \right), \tag{4.211}$$

where $\beta = \frac{3}{2} U K_b T$. Now, we evaluate $\nabla_u V(\mathbf{u})$

$$\nabla_u V(\mathbf{u}) = \frac{\partial V(\varphi)}{\partial \varphi} = 2\beta \cos\varphi \, \sin\varphi, \tag{4.212}$$

which allows us to define the nematic pseudo-rotary velocity

$$\dot{\varphi}_{nem} = -3U D_r \cos(\varphi) \sin(\varphi), \tag{4.213}$$

such that Eq. (4.204), in the absence of flow, reduces to

$$\frac{\partial \Psi}{\partial t} = D_r \frac{\partial^2 \Psi}{\partial \varphi^2} - \frac{\partial}{\partial \varphi} (\dot{\varphi}_{nem} \, \Psi), \tag{4.214}$$

from which we can conclude that Brownian effects induce a tendency towards a random distribution, whereas the nematic term tends to concentrate the orientation distribution more and more in the direction along which most of the molecules are oriented, in the present case, $\varphi = 0$. This tendency can be noticed through simple inspection of Eq. (4.213): there are two equilibrium positions (where the 'nematic" velocity vanishes) at $\varphi = 0$ and $\varphi = \frac{\pi}{2}$. However, that velocity in $\varphi = 0 \pm \varepsilon$ tends to approach the molecule in $\varphi = 0$, whereas in $\varphi = \frac{\pi}{2} \pm \varepsilon$, the velocity moves apart from the molecule that approaches the other equilibrium position $\varphi = 0$, proving that $\varphi = 0$ is the single stable orientation.

4.3.9 Carbon-Nanotube Suspensions: Introducing Aggregation Effects

Carbon nanotubes (CNTs) belong to a relatively new class of fibrous material, with a length scale between that of polymer chain and classical synthetic fibers. Because of their high mechanical strength, low density, and high thermal and electrical conductivity, they can potentially be used for high-performance nanocomposites or nanodevices. Since most applications involve suspending CNTs in a matrix, it is important to understand and model the rheology of CNT suspensions. The rheology of CNTs suspended in different matrices has been studied experimentally by a number of researchers. A significant shear-thinning characteristic for untreated CNTs suspended within an epoxy resin was reported, and CNT aggregates were optically observed.

Although different empirical models, such as the Cross model, the Carreau model and the Krieger–Dougherty model, can be used to describe the evolution of flow curves for CNT suspensions, knowing the microstructure of CNT suspensions is of equal importance, as it is intimately related to the final physical properties of CNT composite materials.

In the context of short-fiber suspensions modeling (previously addressed), a simple orientation model could be used to describe the evolution of steady shear viscosity by coupling the flow kinematics with the fiber orientation. In the first modeling attempt, it was assumed that CNTs can essentially be modeled as short rigid fibers that can rotate and align in a shear flow and that the evolution of the viscosity contribution due to the presence of CNTs depends only on the orientation of the CNTs.

In the simple orientation mesoscopic model, making use of the Fokker–Planck equation and the constitutive law previously introduced, N_p and D_r were the key adjustable parameters to fit into the experimental data. The model was found to be successful in capturing the shear thinning characteristic for some surface-treated CNT suspensions where no optically resolvable CNT aggregate was observed [22].

However, that model, with the use of a single D_r and N_p values, failed to fit the experimental flow curves for untreated CNT suspensions, where the extent of viscosity enhancement was very pronounced at low shear rates. It is clear that a simple orientation explanation was not sufficient to describe the observed viscosity enhancement, and it is believed that the extra contribution to shear viscosity at low shear rates is due to the presence of CNT aggregates, which were experimentally observed in optical microstructure studies.

The simple orientation model could not describe the shear-thinning characteristic for untreated CNT suspensions, and therefore a new model called the Aggregation/Orientation – AO – model was developed. The model essentially considered a hierarchy of CNT aggregate structures in an untreated CNT suspension, where the shear viscosity was controlled not only by CNT orientation, but also by the aggregation state of CNTs in the suspension. The Fokker–Planck description was modified to incorporate aggregation/disaggregation kinetics, and a detailed derivation for the AO model is included in this section.

The aggregation/orientation distribution function in the AO model is written as $\Psi(\mathbf{x}, \mathbf{p}, n, t) : \mathbb{R}^3 \times \mathscr{S} \times [0, 1] \times \mathbb{R}^+ \rightarrow \mathbb{R}^+$, where $n \in [0, 1]$ describes the state of aggregation: $n = 0$ corresponds to CNTs that are free from entanglement and $n = 1$ represents a CNT aggregate network. For a steady and homogeneous flow, the aggregation/orientation distribution reduces to $\Psi(\mathbf{p}, n)$, which describes the fraction of CNTs orientated in the direction $\mathbf{p}$ and belonging to a population n. $\Psi(\mathbf{p}, n)$ now contains information about both CNT orientation and the aggregation state, and the remaining task is to modify the Fokker–Planck equation accordingly.

For an arbitrary population n (where $n \neq 0$ and $n \neq 1$), the population can increase as a result of the aggregation of smaller aggregates ($r < n$) or the disaggregation of larger aggregates ($r > n$). On the other hand, the population n can also decreases because its disaggregation originates fewer entangled aggregates ($r < n$) or the aggregation forms more entangled aggregates ($r > n$). If a constant aggregation velocity (v_c) and a constant disaggregation velocity (v_d) are assumed, the following balance can be written for the population n:

$$\dot{\mathscr{A}}(n) = v_c \int_0^n \Psi(\mathbf{p}, r)\Psi(\mathbf{p}, n-r)\, dr + v_d \int_n^1 \frac{1}{r}\Psi(\mathbf{p}, r)\, dr \\ -v_d \Psi(\mathbf{p}, n) - v_c \Psi(\mathbf{p}, n) \int_0^{1-n} \Psi(\mathbf{p}, r)\, dr, \tag{4.215}$$

where it has been assumed that the desagregation of population $r > n$ produces a uniform distribution of smaller aggregates, justifying the factor $1/r$ in the second integral term.

The resulting Fokker–Planck equation reads as

$$\frac{\partial \Psi}{\partial t} + \nabla_x \cdot (\mathbf{v}\Psi) = -\nabla_p \cdot (\dot{\mathbf{p}}\, \Psi) + \nabla_p \cdot (D_r \nabla_{\mathbf{p}} \Psi) + \dot{\mathscr{A}}(n), \tag{4.216}$$

with the rotary velocity again given by Jeffery's equation for rods (infinite aspect ratio ellipsoids)

$$\dot{\mathbf{p}} = \nabla \mathbf{v} \cdot \mathbf{p} - (\nabla \mathbf{v} : (\mathbf{p} \otimes \mathbf{p}))\, \mathbf{p}. \tag{4.217}$$

The solution of this model allowed for the recovery of the main rheological features of untreated CNT suspensions in [23].

4.3.10 *Kinetic Theory Approach to the Micro Structural Theory of Passive Mixing*

Many chemical engineering processes benefit from good bulk mixing. In such processes, determination of the mixing rate is important in terms of understanding and predicting the mixing time. The understanding allows us to define new flows and the associated processes maximizing the mixing rate, but one could also try to optimize other parameters related to the microstructure describing the morphology, characteristic length, shape and orientation.

One possible way to quantify the mixing rate can be derived from consideration of only one of the basic mechanisms of mixing processes: the increase of the material interface due to the fluid mechanics in absence of interfacial tension and molecular diffusion. The molecular diffusion leads to smooth concentration gradients across the interface. However, this mechanism only becomes significant when the interface has increased significantly. It is therefore expected that the overall mixing rate will be closely linked to the rate of mechanical stretching of the interfacial area.

The approaches quantifying the mixing from the increase of the material interface have two important drawbacks: (i) the first one is associated with the difficulty of introducing other additional physics, such as the one related to the surface tension, and (ii) sometimes the microstructure description needs additional information (morphology, characteristic length, shape and orientation, etc.) that the area of the interface evolution cannot provide.

In passive mixing, interfacial energy is negligible and the two phases have identical viscosities. Moreover, the global velocity field can be found independently of the microstructure and then used to evolve the mixture structure, described with some area tensor, very rich from the morphological and microstructural points of view.

In this section, we summarize the kinetic theory approach proposed in our former work [24] based on the use of the area tensor considered [25].

4.3.10.1 Morphological Description of Microstructured Fluids

Let Ω be the domain in which the flow problem is defined. Points in Ω will be referred by $\mathbf{x}$, which is a vector in 2D or 3D. In order to quantify the morphology at any point $\mathbf{x} \in \Omega$, a microscopic representative volume $V(\mathbf{x})$ is considered to be centered at that point. This volume turns out to be small in respect to the macroscopic scale (related to the variation of the velocity field in Ω), but large enough in respect to the characteristic size of the microstructure. Let $S(\mathbf{x})$ be the interface within $V(\mathbf{x})$. The second order area tensor $\mathbf{A}$ is then defined as:

$$\mathbf{A}(\mathbf{x}) = \frac{1}{V(\mathbf{x})} \int_{S(x)} \mathbf{n} \otimes \mathbf{n} \, dS, \tag{4.218}$$

where $\mathbf{n}$ represents the unit vector defined on the interface $S(\mathbf{x})$, assuming that it is pointing towards the continuous phase (the discrete one is the one with the lower volume fraction). This area tensor is symmetric and has different appealing properties:

- The first property concerns its trace, which we symbolize with $\mathrm{Tr}(\mathbf{A}(\mathbf{x}))$, which, taking into account the normality of $\mathbf{n}$, results in

$$\mathrm{Tr}(\mathbf{A}(\mathbf{x})) = \frac{1}{V(\mathbf{x})} \int_{S(\mathbf{x})} (\mathbf{n}_1^2 + \mathbf{n}_2^2 + \mathbf{n}_3^2) \, dS = \frac{1}{V(\mathbf{x})} \int_{S(\mathbf{x})} dS = \frac{\mathbf{S}(\mathbf{x})}{\mathbf{V}(\mathbf{x})} = S_v(\mathbf{x}), \tag{4.219}$$

  where $S_v(\mathbf{x})$ represents the specific surface related to point $\mathbf{x}$, and whose maximization is usually searched in mixing processes.

- If we define the volume fraction of the disperse phase as ϕ, then we can define a characteristic length of the microstructure at point $\mathbf{x}$ from the volume of the discrete phase $V_d(\mathbf{x}) = \phi V(\mathbf{x})$:

$$L(\mathbf{x}) \equiv \frac{V_d(\mathbf{x})}{S(\mathbf{x})} = \frac{\phi V(\mathbf{x})}{S(\mathbf{x})} = \frac{\phi}{S_v(\mathbf{x})} = \frac{\phi}{\mathrm{Tr}(\mathbf{A}(\mathbf{x}))}. \tag{4.220}$$

- The microstructure's shape and orientation can be easily deduced from the normalized area tensor $\tilde{\mathbf{A}}(\mathbf{x})$, defined as

$$\tilde{\mathbf{A}}(\mathbf{x}) = \frac{\mathbf{A}(\mathbf{x})}{\mathrm{Tr}(\mathbf{A}(\mathbf{x}))}, \tag{4.221}$$

  and represented by an ellipsoid. The eigenvalues of $\tilde{\mathbf{A}}$ allow us to compute the length of the ellipsoid axes, since their orientation is given by the associated eigenvectors.

The interest in computing the area tensor, from a given initial condition, at each point of the flow domain and at each time has been justified. However, its evaluation requires establishment of the partial differential equation governing its evolution. In what follows, we establish this evolution equation and the closure issue that its solution implies.

To derive the evolution equation of $\mathbf{A}(\mathbf{x})$, we consider the time derivative of vector $\mathbf{n}$ and the surface element dS (standard results in continuum mechanics)

$$\frac{d\mathbf{n}}{dt} = \dot{\mathbf{n}} = -(\nabla\mathbf{v})^{\mathrm{T}}\mathbf{n} + (\nabla\mathbf{v} : (\mathbf{n} \otimes \mathbf{n}))\,\mathbf{n} \tag{4.222}$$

and

$$\frac{dS}{dt} = -(\nabla\mathbf{v} : (\mathbf{n} \otimes \mathbf{n}))\,dS. \tag{4.223}$$

Introducing Eqs. (4.222) and (4.223) into the time derivative of Eq. (4.218), it results that

$$\frac{d\mathbf{A}}{dt} = -(\nabla\mathbf{v})^{\mathbf{T}} \cdot \mathbf{A} \;-\; \mathbf{A} \cdot \nabla\mathbf{v} + \nabla\mathbf{v} : \mathscr{A}, \tag{4.224}$$

where the dependence of $\mathbf{A}(\mathbf{x})$ on $\mathbf{x}$ has been omitted for the sake of clarity, and where $\mathscr{A}$ denotes the fourth order area tensor defined by

$$\mathscr{A}(\mathbf{x}) = \frac{1}{V(\mathbf{x})} \int_{S(\mathbf{x})} \mathbf{n} \otimes \mathbf{n} \otimes \mathbf{n} \otimes \mathbf{n}\, dS. \tag{4.225}$$

The main difficulty in solving Eq. (4.224) is precisely related to that fourth order area tensor, as was the case in suspensions composed of rods. Thus, a closure relation expressing that tensor as a function of the second order area tensor is needed. In absence of a general exact closure relation, any closure proposal must be checked carefully, because its impact on the computed solution is a priori unpredictable.

The formalism associated with the evolution of the area tensor has other advantages. One of them is the simplicity of incorporating other physical effects, for example, the surface tension. One could expect that in absence of strain rate, the disperse microstructure evolves towards an isotropic state composed of coalescent microspheres, minimizing the specific area. Thus, Doi and Ohta [26] proposed introducing into Eq. (4.224) a source term accounting for such effects

$$\frac{d\mathbf{A}}{dt} = -(\nabla\mathbf{v})^{\mathrm{T}} \cdot \mathbf{A} - \mathbf{A} \cdot \nabla\mathbf{v} + \nabla\mathbf{v} : \mathscr{A} - a\frac{\sigma}{\eta} S_v^2 \left(\left(\tilde{\mathbf{A}} - \frac{\mathbf{I}}{3} \right) + b\frac{\mathbf{I}}{3} \right), \quad (4.226)$$

where a and b are two material parameters, σ is the surface tension, η the fluid viscosity (we are assuming that both fluids have the same viscosity) and $\mathbf{I}$ the unit tensor. Thus, in the absence of strain rate, Eq. (4.226) reduces to

$$\frac{d\mathbf{A}}{dt} = -a\frac{\sigma}{\eta} S_v^2 \left(\left(\tilde{\mathbf{A}} - \frac{\mathbf{I}}{3} \right) + b\frac{\mathbf{I}}{3} \right), \quad (4.227)$$

proving that the microstructure evolves towards an isotropic state induced by the presence of the term $\tilde{\mathbf{A}} - \frac{\mathbf{I}}{3}$, in which the interface area reduces due to the presence of $-\frac{ab\sigma}{\eta} S_v^2 \frac{\mathbf{I}}{3}$. Even if this model predicts a null long-time specific area, a fact that motivated different corrections, we will consider the Doi-Ohta model in the kinetic theory approach that follows.

4.3.10.2 Kinetic Theory Description of Passive Mixing

We define the area distribution function $\Psi(\mathbf{x}, \mathbf{n}, t) : \mathbb{R}^3 \times \mathscr{S} \times \mathbb{R}^+ \rightarrow \mathbb{R}^+$, given at each point in the physical domain $\mathbf{x} \in \Omega$ and for any time t, the specific surface $S_v(\mathbf{x})$ orientated in the direction $\mathbf{n}$. Thus, the area tensor can be defined from this area distribution function according to

$$\mathbf{A}(\mathbf{x}, t) = \int_{\mathscr{S}} \mathbf{n} \otimes \mathbf{n} \, \Psi(\mathbf{x}, \mathbf{n}, t) \, d\mathbf{n}. \quad (4.228)$$

The trace of the previous equation reads as expected:

$$\mathrm{Tr}(\mathbf{A}(\mathbf{x}, t)) = S_v(\mathbf{x}, t) = \int_{\mathscr{S}} \mathrm{Tr}(\mathbf{n} \otimes \mathbf{n}) \, \Psi(\mathbf{x}, \mathbf{n}, t) \, d\mathbf{n} = \int_{\mathscr{S}} \Psi(\mathbf{x}, \mathbf{n}, t) \, d\mathbf{n}. \quad (4.229)$$

The expression of the time derivative of the distribution function reads as

$$\frac{\partial \Psi}{\partial t} + \nabla_x \cdot (\mathbf{v}\Psi) = -\nabla_n \cdot (\dot{\mathbf{n}}\Psi) - (\nabla \mathbf{v} : (\mathbf{n} \otimes \mathbf{n}))\,\Psi. \tag{4.230}$$

Equation (4.230) is close to the Fokker–Planck equation usually employed to describe the microstructure in the rods suspensions considered in the previous sections. However, in the present case, because this function is not subjected to a normality condition (Ψ is not a probability density function), there is an additional term (last term in the right member of Eq. (4.230)) taking into account the growth of the interface area induced by its stretching.

If Eq. (4.230) is solved instead of Eq. (4.224) (the latter requires the choice of a closure relation for the fourth order orientation tensor), the area tensor can be computed from Eq. (4.228) without the necessity of introducing any closure relation.

Finally, other physical effects, and in particular, the one related to the surface tension, can be easily introduced in the kinetic theory formalism. As before, we consider the expression of the time derivative of the area tensor (which in the kinetic theory framework is given by Eq. (4.228)), looking for the time derivative of the distribution function Ψ leading to the expression (4.226).

It is easy to prove that the associated kinetic model is given by:

$$\frac{d\Psi}{dt} = -\frac{\partial}{\partial \mathbf{n}}(\dot{n}\,\Psi) - \Big(\nabla \mathbf{v} : (\mathbf{n} \otimes \mathbf{n})\Big)\Psi + \frac{\partial}{\partial \mathbf{n}}\left(\mathscr{D}\frac{\partial \Psi}{\partial \mathbf{n}}\right) + \mathscr{F}, \tag{4.231}$$

where the diffusion coefficient and the source terms result in 2D

$$\mathscr{D} = \frac{a}{4}\frac{\sigma}{\eta} S_v = \frac{a}{4}\frac{\sigma}{\eta}\left(\int_{\mathscr{S}} \Psi(\mathbf{n})\, d\mathbf{n}\right) \tag{4.232}$$

and

$$\mathscr{F} = \frac{a}{4\pi}\frac{\sigma}{\eta}(S_v)^2 b = -\frac{ab}{4\pi}\frac{\sigma}{\eta}\left(\int_{\mathscr{S}} \Psi(\mathbf{n})\, d\mathbf{n}\right)^2. \tag{4.233}$$

4.4 The Chemical Master Equation

Simulating, for example, the behavior of gene regulatory networks is a formidable task for several reasons. At this level of description, only a few molecules (maybe dozens or hundreds) of each species involved in the regulation process are present, and this fact limits the possibility of considering the process to be deterministic, as is done very often in most chemical applications. Here, the concept of concentration of the species does not make sense. On the contrary, under some weak hypotheses, the system can be considered as Markovian (memory-less), and can consequently be modeled by the so-called Chemical Master Equation – CME–, which, is in fact, nothing more than a set of ordinary differential equations stating the conservation of the probability distribution function – pdf – P in time.

We consider n different chemical species $\mathscr{S}_i$, $i = 1, \ldots, n$, each one having $K_i + 1$ possible number of individuals $\#s_i \in (0, 1, \ldots, K_i)$, $i = 1, \ldots, n$, and a set of m reactions R_j, $j = 1, \ldots, m$ with propensity a_j encoding the reaction rate, expressed by

$$\chi_{j,1}^{-} s_1^{-} \mathscr{S}_1 + \ldots + \chi_{j,n}^{-} s_n^{-} \mathscr{S}_n \xrightarrow{a_j} \chi_{j,1}^{+} s_1^{+} \mathscr{S}_1 + \ldots + \chi_{j,n}^{+} s_n^{+} \mathscr{S}_n; \quad j = 1, \ldots, m, \tag{4.234}$$

with $\chi_{j,i}$ controlling the appearance of specie S_i in reaction j. For that purpose, χ is a boolean variable taking values 0 or 1.

The system state is defined from $z = (\#s_1, \ldots, \#s_n)$. Thus, reaction j transforms the state $\hat{z} = (\#s_1^-, \ldots, \#s_n^-)$ into z, with $z = \hat{z} + v_j$, where $v_j = (\#s_1^+ - \#s_1^-, \ldots, \#s_n^+ - \#s_n^-)$ contains the j-reaction stoichiometry.

As a memoryless random walk, each action corresponds to a Markov equation, the so-called Chemical Master Equation – CME – which tracks the net change in the probability $P(z, t)$

$$\frac{dP(z,t)}{dt} = \sum_{j=1}^{m} a_j(z - v_j, t) P(z - v_j, t) - a_j(z, t) P(z, t), \tag{4.235}$$

with the propensity a_j depending on the system state and time.

Equation 4.235 constitutes a system of linear differential equations that can be integrated from a given initial condition. What is challenging is that it is defined in a space with as many dimensions as the number of different species involved in the regulatory network. If we consider N different species, present at a number n of copies, the number of different possible states of the system is n^N. This number can take the astronomical value of 10^{6000} if we consider certain types of proteins, for instance.

To overcome this difficulty, most researches employ Monte Carlo-like algorithms (stochastic simulations). However, Monte Carlo techniques need as many individual realizations of the problem as possible, leading to excessive time-consuming simulations, together with great variance in the results. In [27], the authors consider the proper generalized decomposition for alleviating the curse of dimensionality. Then, in [28], empirical closures were derived and used within a moment-based description.

4.4.1 *Moment Based Descriptions*

The issue related to the curse of dimensionality, in many other disciplines as previously described (e.g., suspensions), motivated the replacement of the pdf with some of its moments, since they very often suffice for having a rich enough view of the dynamics of the systems. The use of moment-based descriptions was of major interest in different areas of statistical mechanics, and its consideration as an alternative to the discretization of the CME is increasing.

A moment represents the expected value of a random variable, z, raised to a certain power. An "expectation" is a specifically defined function in statistics, $E[f(z)] = \int f(z)P(z)dz$ when in continuous spaces or $\sum f(z)P(z)$ in discrete spaces. In general, we can talk about the ith moment as

$$\mu_i(t) = E\left[z^i\right] = \sum_{z=0}^{\infty} P(z,t)z^i. \tag{4.236}$$

A probability distribution is uniquely defined by its full set of moments. Having access to these moments could eliminate the need to solve for the full distribution, depending on what information would be considered important. A special function, called the Moment Generating Function $M(\theta,t)$, is specifically intended for this purpose:

$$M(\theta,t) = \sum_{z=0}^{\infty} e^{\theta z} P(z,t). \tag{4.237}$$

By taking the Taylor expansion of $e^{\theta z} = 1 + \frac{(\theta z)^1}{1!} + \frac{(\theta z)^2}{2!} + \frac{(\theta z)^3}{3!} + \cdots$, we can see the moments emerging from this function, the i-th moment associated with the i-th power of θ

$$M(\theta,t) = \mu_0(t) + \frac{\mu_1(t)\theta}{1!} + \frac{\mu_2(t)\theta^2}{2!} + \cdots = \sum_{z=0}^{\infty} \frac{\mu_z(t)\theta^z}{z!}. \tag{4.238}$$

The following equations will be used extensively in the subsequent derivation, so it will be useful to define them before we proceed:

$$\begin{cases} M(\theta,t) = \sum_{z=0}^{\infty} e^{\theta z} P(z,t) = \sum_{z=0}^{\infty} \frac{\mu_z(t)\theta^z}{z!} \\ \frac{\partial M(\theta,t)}{\partial t} = \sum_{z=0}^{\infty} e^{\theta z} \frac{\partial P(z,t)}{\partial t} \\ \frac{\partial^i M(\theta,t)}{\partial \theta^i} = \sum_{z=0}^{\infty} e^{\theta z} P(z,t) z^i = \sum_{z=0}^{\infty} \frac{\mu_z(t)\theta^{z-i}}{(z-i)!} \end{cases}. \tag{4.239}$$

4.4.2 From the Chemical Master Equation to Moment-Based Descriptions

Since we will only consider the structure of the Chemical Master Equation, we would like to derive a general version of the Moment Generating Function that can be used for any system. The CME for l reactions with stoichiometric change v_l is:

$$\frac{\partial P(z,t|z_0,t_0)}{\partial t} = \sum_l a_l(z-v_l)P(z-v_l,t) - a_l(z)P(z,t). \tag{4.240}$$

As we will see later on, the kind of rate laws associated with the system dramatically impacts the complexity of the overall problem. We consider the simplest case of kinetic mass action laws. An example of a mass action rate law is $a_l(z) = \frac{\lambda z_1(z_1-1)}{2} = \frac{\lambda}{2}z_1^2 - \frac{\lambda}{2}z_1 = \sum_i c_{l,i}a\prime_{l,i}$, where the law can be rewritten as a sum of coefficients $c_{l,i}$ and variables $a\prime_{l,i}$. This expanded, polynomial form will be exploited in our derivation.

Since we would like to talk about moments of the CME rather than probabilities, our first priority is to write this equation in terms of M, rather than in terms of P. We multiply both sides by $e^{\theta z}$ and sum over all possible values of z

$$\sum_{z=0}^{\infty} e^{\theta z}\frac{\partial P(z,t)}{\partial t} = \sum_{z=0}^{\infty}\sum_l e^{\theta z}a_l(z-v_l)P(z-v_l,t) - e^{\theta z}a_l(z)P(z,t), \tag{4.241}$$

which taking into account the previous definitions, results in

$$\begin{aligned}\frac{\partial M(\theta,t)}{\partial t} &= \sum_{z=0}^{\infty}\sum_l\left(\sum_i c_{l,i}a\prime_{l,i}(z-v_l)e^{\theta z}P(z-v_l,t) - \sum_i c_{l,i}a\prime_{l,i}(z)e^{\theta z}P(z,t)\right)\\ &= \sum_{z=0}^{\infty}\sum_l\left(\sum_i c_{l,i}a\prime_{l,i}(z-v_l)e^{\theta(z-v_l)}e^{\theta(v_l)}P(z-v_l,t) - \sum_i c_{l,i}a\prime_{l,i}(z)e^{\theta z}P(z,t)\right)\\ &= \sum_l\left(\sum_i c_{l,i}\frac{\partial^i M}{\partial\theta^i}e^{\theta(v_l)} - \sum_i c_{l,i}\frac{\partial^i M}{\partial\theta^i}\right) = \sum_l\sum_i c_{l,i}\frac{\partial^i M}{\partial\theta^i}\left(e^{\theta(v_l)}-1\right).\end{aligned} \tag{4.242}$$

Now, we can take the second definition of $\frac{\partial^i M}{\partial\theta^i}$ and expand $e^{\theta(v_l)}$ into its Taylor series. Note that the summation now begins at $j = i$. When $j < i$, the index will be out of bounds and will not correspond to any physical state

$$\frac{\partial M(\theta,t)}{\partial t}=\sum_l\sum_i c_{l,i}\frac{\partial^i M}{\partial\theta^i}\left(e^{\theta(\nu_l)}-1\right)$$

$$=\sum_l\sum_i c_{l,i}\sum_{j=i}^{\infty}\frac{\mu_j(t)\theta^{j-i}}{(j-i)!}\left(\sum_{k=0}^{\infty}\frac{(\theta\nu_l)^k}{k!}-1\right). \tag{4.243}$$

Remember that the initial goal was to isolate the coefficients of θ^n in order to obtain the nth moments:

$$\frac{\partial M(\theta,t)}{\partial t}=\sum_l\sum_i c_{l,i}\sum_{j=i}^{\infty}\frac{\mu_j(t)\theta^{j-i}}{(j-i)!}\left(\sum_{k=0}^{\infty}\frac{(\theta\nu_l)^k}{k!}-1\right)$$

$$=\sum_l\sum_i c_{l,i}\left(\frac{\mu_i}{0!}+\frac{\mu_{i+1}\theta}{1!}+\frac{\mu_{i+2}\theta^2}{2!}+\cdots\right)\left(\frac{\nu_l\theta}{1!}+\frac{(\nu_l\theta)^2}{2!}+\frac{(\nu_l\theta)^3}{3!}+\cdots\right)$$

$$=\sum_l\sum_i c_{l,i}\left(\left[\frac{\mu_i}{0!}\frac{\nu_l}{1!}\right]\theta+\left[\frac{\mu_i}{0!}\frac{\nu_l{}^2}{2!}+\frac{\mu_{i+1}}{1!}\frac{\nu_l}{1!}\right]\theta^2+\cdots\right)$$

$$=\sum_l\sum_i c_{l,i}\sum_{n=0}^{\infty}\theta^n\sum_{k=1}^{n}\mu_{i+(n-k)}\frac{1}{k!(n-k)!}. \tag{4.244}$$

Our next step will be to isolate only the coefficients of θ, in order to achieve a form in which we are creating ODE's of μ, rather than $M(\theta,t)$. Since $\frac{\partial M(\theta,t)}{\partial t}=\sum_n\frac{\partial\mu_n}{\partial t}\frac{1}{n!}\theta^n$, we will have to multiply both sides by $n!$ in order to isolate μ. Thus, it finally results that

$$\frac{\partial\mu_n(t)}{\partial t}=\sum_l\sum_i c_{l,i}\sum_{k=1}^{n}\nu_l^k\mu_{i+(n-k)}\frac{n!}{k!(n-k)!}. \tag{4.245}$$

It is easy to note from the previous expression (4.245) that the equation that governs the time evolution of the moments up to a certain order implies, in general, higher order moments, and thus, before solving all of them, higher order moments must be written as a combination of those involved in the time evolution equations considered. These relations have, in most cases, an approximate character and are known as closure relations.

References

1. E. Schrödinger, What is life? The physical aspect of the living cell. Dublin Institute for Advanced Studies at Trinity College, Dublin (1944)
2. F. Chinesta, E. Abisset, A. Ammar, E. Cueto, Efficient numerical solution of continuous mesoscale models of complex fluids involving the Boltzmann and Fokker–Planck equations. Commun. Comput. Phys. **17**(4), 975–1006 (2015)
3. S. Succi, *The Lattice Boltzmann Equation for Fluid Dynamics and Beyond* (Clarendon Press, Oxford, 2001)
4. G.B. Jeffery, The motion of ellipsoidal particles immersed in a viscous fluid. Proc. R. Soc. Lond. A **102**, 161–179 (1922)
5. C. Binetruy, F. Chinesta, R. Keunings, *Flows in Polymers, Reinforced Polymers and Composites. A Multiscale Approach*, Springerbriefs (Springer, Berlin, 2015)
6. C.V. Chaubal, A. Srinivasan, O. Egecioglu, L.G. Leal, Smoothed particle hydrodynamics techniques for the solution of kinetic theory problems. J. Non-Newtonian Fluid Mech. **70**, 125–154 (1997)
7. F. Chinesta, G. Chaidron, A. Poitou, On the solution of the Fokker–Planck equation in steady recirculating flows involving short fibre suspensions. J. Non-Newtonian Fluid Mech. **113**, 97–125 (2003)
8. A. Ammar, F. Chinesta, A particle strategy for solving the Fokker–Planck equation governing the fibre orientation distribution in steady recirculating flows involving short fibre suspensions, in *Lectures Notes on Computational Science and Engineering*, vol. 43 (Springer, Berlin, 2005), pp. 1–16
9. F. Chinesta, A. Ammar, A. Falco, M. Laso, On the reduction of stochastic kinetic theory models of complex fluids. Model. Simul. Mater. Sci. Eng. **15**, 639–652 (2007)
10. H.C. Ottinger, *Stochastic Processes in Polymeric Fluids* (Springer, Berlin, 1996)
11. A. Ammar, B. Mokdad, F. Chinesta, R. Keunings, A new family of solvers for some classes of multidimensional partial differential equations encountered in kinetic theory modeling of complex fluids. J. Non-Newtonian Fluid Mech. **139**, 153–176 (2006)
12. A. Ammar, B. Mokdad, F. Chinesta, R. Keunings, A new family of solvers for some classes of multidimensional partial differential equations encountered in kinetic theory modeling of complex fluids. Part II. J. Non-Newtonian Fluid Mech. **144**, 98–121 (2007)
13. F. Chinesta, A. Ammar, A. Leygue, R. Keunings, An overview of the proper generalized decomposition with applications in computational rheology. J. Non-Newtonian Fluid Mech. **166**, 578–592 (2011)
14. F. Chinesta, R. Keunings, A. Leygue, *The Proper Generalized Decomposition for Advanced Numerical Simulations: A Primer*, SpringerBriefs in Applied Science and Technology (Springer, Berlin, 2014)
15. E. Abisset-Chavanne, F. Chinesta, J. Ferec, G. Ausias, R. Keunings, On the multiscale description of dilute suspensions of non-Brownian rigid clusters composed of rods. J. Non-Newtonian Fluid Mech. **222**, 34–44 (2015)
16. E. Abisset-Chavanne, R. Mezher, S. Le Corre, A. Ammar, F. Chinesta, Kinetic theory microstructure modeling in concentrated suspensions. Entropy **15**, 2805–2832 (2013)
17. R. Tanner, *Engineering Rheology* (Oxford University Press, Oxford, 1985)
18. R.B. Bird, C.F. Curtiss, R.C. Armstrong, O. Hassager, *Dynamic of Polymeric Liquids* (Wiley, New York, 1987)
19. M. Doi, S.F. Edwards, *The Theory of Polymer Dynamics* (Clarendon Press, Oxford, 1987)
20. P.G. de Gennes, Repation of a polymer chain in the presence of fixed obstacles. J. Chem. Phys. **55**(2), 572–579 (1971)
21. M. Doi, S.F. Edwards, Dynamics of rod-like macromolecules in concentrated solution. J. Chem. Soc. Faraday Trans. **74**, 560–570 (1978)
22. A. Ma, F. Chinesta, M. Mackley, The rheology and modelling of chemically treated carbon nanotube suspensions. J. Rheol. **53**(3), 547–573 (2009)

23. A. Ma, F. Chinesta, A. Ammar, M. Mackley, Rheological modelling of carbon nanotube aggregate suspensions. J. Rheol. **52**(6), 1311–1330 (2008)
24. F. Chinesta, M. Mackley, Microstructure evolution during liquid-liquid laminar mixing: a kinetic theory approach. Int. J. Mater. Form. **1**, 47–55 (2008)
25. E.D. Wetzel, C.L. Tucker III, Area tensors for modeling microstructure during laminar liquid-liquid mixing. Int. J. Multiph. Flow **25**, 35–61 (1999)
26. M. Doi, T. Ohta, Dynamics and rheology of complex interfaces. J. Chem. Phys. **95**, 1242–1248 (1991)
27. A. Ammar, E. Cueto, F. Chinesta, Reduction of the chemical master equation for gene regulatory networks using proper generalized decompositions. Int. J. Numer. Methods in Biomed. Eng. **28**(9), 960–973 (2012)
28. A. Ammar, M. Magnin, O. Roux, E. Cueto, F. Chinesta, Chemical master equation empirical moment closure. Biol. Syst. **5/1**, 1000155 (2016)

Printed in the United States
By Bookmasters